Tuning Accel/DFI 6.0 Programmable Fuel Injection

• Understanding Electronic Engine Management • Fuel Injector Dynamics • Using CalMap Software • Tuning Tips for Race, Street, Marine, and Off-Road • Covers GM, Ford, Mopar, & Honda

Ray T. Bohacz

HPBOOKS

HPBooks
are published by
The Berkley Publishing Group
A division of Penguin Putnam Inc.
375 Hudson Street
New York, New York 10014

First edition: July 2003
ISBN: 1-55788-413-7

10 9 8 7 6 5 4 3 2 1

This book has been catalogued with the Library of Congress

Book design and production by Michael Lutfy
Cover photo by Ray T. Bohacz
Cover design by Bird Studios
Interior photos and illustrations by author and courtesy ACCEL

NOTICE: The information in this book is true and complete to the best of our knowledge. All recommendations on parts and procedures are made without any guarantees on the part of the author or the publisher. Tampering with, altering, modifying or removing any emissions-control device is a violation of federal law. Author and publisher disclaim all liability incurred in connection with the use of this information.

CONTENTS

ACKNOWLEDGMENTS

Dedication

Dedicated to my precious wife and forever soulmate, Charlotte; my dog, Sparky, all of the men and women in our armed forces; and President and Laura Bush.

About the Author

Ray T. Bohacz's interests have always revolved around mechanical apparatus, but he admits that his true love is engines. Ray graduated high school by age 16 and was the youngest student on the college campus. He began studying electrical engineering, and the young hot-rodder became disenchanted when his professor couldn't explain how the then-new automotive electronic ignitions functioned.

With the return of Detroit performance in 1985, Ray could not be left on the sidelines. He opened a speed shop that specialized in tuning and modifications to EFI cars. In 1993, he was the first to modify a fourth-generation Camaro. That car ran 11.70 at 119 mph and took second place in the *Car Craft Magazine* Real Street Eliminator contest in 1994.

Disappointed that the magazines were not giving readers any real technical information, he decided to try his hand at writing. His first byline appeared in 1995 in *Vette* magazine, and he is a regular contributor to *Hot Rod, Popular Hot Rodding, GM High Tech Performance, High Performance Pontiac, Corvette Fever, High Performance Mopar, Hemmings Rods and Performance, Muscle Car Review, Specialty Interest Auto,* and *Mustangs Illustrated*, among many others. He also created the ACCEL EMIC training manual and the course curriculum taught at the University of Northwestern Ohio in Lima. Ray is a member of the Society of Automotive Engineers, the American Society of Materials, and the International Motoring Press Association.

When not writing or acting as a consultant to the automotive industry, Ray enjoys spending time with his wife, Charlotte, and a fully restored 1940 Ford 9N farm tractor; or working the fields with the help of his dog, Sparky, and a 2002 Ford New Holland TC-29 FWD tractor

INTRODUCTION

This book was created to fill a void due to the lack of published information concerning the use and tuning of the ACCEL/DFI programmable electronic fuel injection. The product first appeared in limited numbers around 1986 and was a fuel-control ECU (engine control unit) only. Shortly thereafter, a companion product marketed under the name Power Processor was sold alongside. The Power Processor was unique in that it was not a stand-alone ECU, but what eventually became known as a piggyback unit. It worked in conjunction with a General Motors 1986-89 Firebird/Camaro/Corvette processor. Simply put, the Power Processor allowed the limited ability to alter the fuel table and wide-open throttle timing map only. The fuel-control ECU was known as the Version 5.0 FM, designating fuel management. Versions 1 through 4 were prototypes and were never marketed.

Version 5.0 FM offered full control of the fuel table, along with ancillary functions such as idle air control, warm-up enrichment, cranking fuel, etc. Both the Power Processor and the Version 5.0 FM were the industry's first engine management systems to allow "on-the-fly" tuning. This means you can make calibration changes with a PC while the engine is running. Other manufacturers and even most original-equipment ECUs require tuners to download the tables from the management system to a PC, create a file, edit the file offline from the ECU, and then reboot the table in the car's electronics. This is a very timely and complicated procedure. In addition, it would take many steps and a few minutes to alter only one point in the calibration. With the ACCEL/DFI system, a change can be made and a determination as to its effect on the engine qualified almost immediately.

This book is dedicated to the Version 6.0 Spark/Fuel Management, or SFM system. It first appeared on the market in the spring of 1991 and replaced the Version 5.0 FM design. The Version 6.0 offered fuel and total timing control on certain applications. GM electronic spark timing ignition was compatible and allowed mapping of the complete spark advance curve. Shortly after Version 6.0's introduction, many alterations on the same ECU were offered, allowing fuel and timing control on the Ford EEC-IV Thick Film Ignition, the GM LT1 Opti-Spark, and the Buick Grand National Turbo V-6 distributor-less ignition. To allow integration with other engines, an inductive pickup ECU was created from the same basic architecture. Through the use of an aftermarket capacitive-discharge ignition amplifier such as an ACCEL 300+, MSD, Mallory, etc., complete spark control and fuel control were available on any engine. This opened the door to the acceptance of the ACCEL/DFI in all forms of motorsports, as well as for unique applications that included industrial and truck engines, and the then-not-yet-burgeoning sport compact market.

Just as this book was being completed, ACCEL introduced a new and anxiously awaited Version 7.0 ECU to replace the long-running Version 6.0. Production on the Version 6.0 lasted more than 11 years, making it one of the most widely used and popular aftermarket ECU in history. Sadly, ACCEL never offered any meaningful instructions or guide for using the Version 6.0 ECU, thus leaving a terrible void in knowledge for the end user. That became the impetus for creating this text, since there is a great pent-up demand for in-depth information about the intricacies of this system.

To become a successful tuner, you actually need to be a calibrator. This creates the concern that there are two distinct sciences at work: knowledge of how an engine and a fuel-injection system function, and the operation and understanding of the CalMap software. Knowledge of one without the other is meaningless. To this cause, tuning ACCEL/DFI works under the assumption that the basic elements of combustion and the four strokes of an engine are understood by the reader. It is strongly suggested that you not overlook or make light of these basic theories. When writing a calibration with the ACCEL/DFI, a lack of engine theory not only will make the task almost impossible, but can lead to severe damage to the engine or the driver of the vehicle. Often an enthusiast will assume he knows how an engine works, but in reality he has only, at best, a superficial understanding of the theories involved. Many excellent books are available that can explain the basics of the four-stroke cycle and should be studied as part of your calibration career.

Taking a systematic approach, this text is written in a way that will bring you slowly into becoming a calibrator. What's nice about this method is that it educates the reader in the basic functions of electronic fuel injection, and this knowledge can be applied to almost every style or brand of fuel-injection system offered to date. Much effort was taken for this not to be a software guide, but to include engine-management review, fuel-flow dynamics, and emissions compliance, along with ancillary areas such as intake manifold selection. Not until Chapter 6 is the CalMap software discussed in-depth. To help the reader, and to be a self-sufficient resource, other features were included. One chapter is dedicated to the tuning and specifics of working with GM, Ford, and Honda engines and vehicles. This represents the lion's share of ACCEL/DFI installations, and the differences in the engines demanded special attention. The book also includes a glossary of relevant terms and allows you to build your knowledge beyond the subject matter.

The CalMap software is user-friendly, but to truly learn it, not only will you need to work with it, but you might want to read Chapters 6 and 7 more than once. There's simply too much information to be absorbed in one reading, and as you work with the system, new obstacles and questions will arise. This book was designed to be a tool and a companion early on in your tuning career. It was not intended to be glanced over once or twice and placed on a shelf. Before working with the system, reread and study the calibration tables and their function, and how they interact with the engine and other parts of the software. As your skill sets improve and you have accomplished more successful calibrations, the text will assume the role of a resource, readily at hand if required.

As the old saying goes, "There's more than one way to skin a cat." That saying also can be applied to creating a calibration with the ACCEL/DFI system. The end result should always be a proper-running engine, and this can usually be accomplished in many different ways. From my experience working with this system and teaching it to ACCEL dealers, I have discovered that almost everyone is looking for a magic pill, the right key in the software to press. Well, let me make this perfectly clear: The only magic pill is knowledge, and when you posses it, you will know which is the proper table or value to enter. Good luck, and enjoy the beginning of your journey as a calibration engineer.

Ray T. Bohacz
February 2003

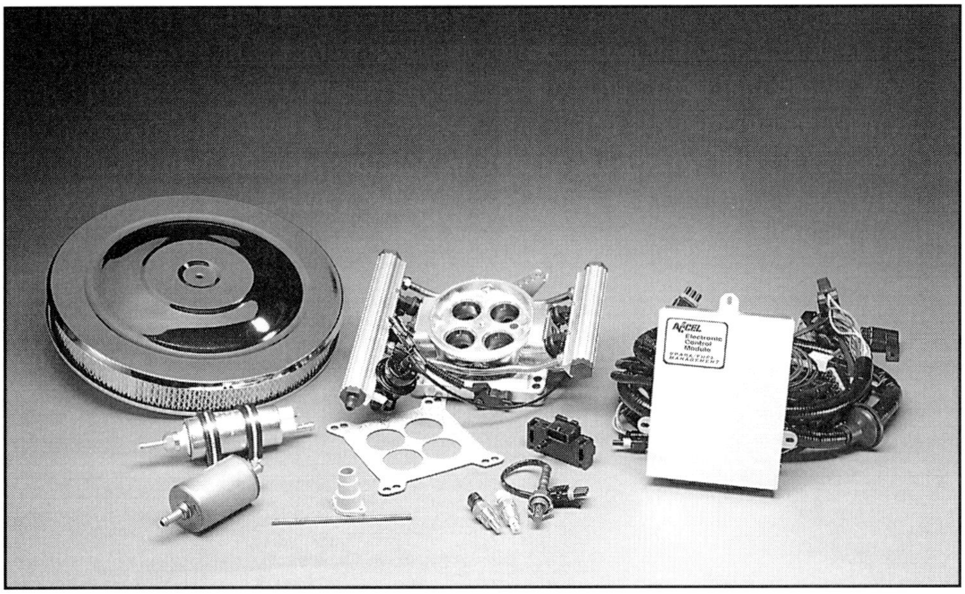

1

Engine Management Basics

Why Fuel Injection?

In racing, fuel injection is associated with the ultimate in performance, but the reason behind its proliferation onto mainstream automobiles is for quite the opposite purpose. By the early 1980s, increasingly stricter emissions standards and the EPA's implementation of a minimum fuel economy standard meant that the days of the carburetor were numbered. Without electronic engine controls, it simply was not possible to accurately command all the necessary parameters to remain compliant; finite control of the fuel and ignition systems became necessary. The advent of cold-start emissions testing made apparent the need to decrease catalytic converter light-off time, and made critical the cold start and intermediate fueling.

Even though fuel injection made its debut in America in the 1950s with General Motors using it on certain Chevrolet and Pontiac models, today's injection bears no resemblance to that rudimentary beginning. The Europeans, working in conjunction with Bosch, were the pioneers of modern fuel injection. The high cost of fuel in Europe and the taxation of large powerplants forced the development of fuel-stingy, small-displacement engines. With no need to respond to the same conditions, American manufacturers were late to the injection game, and in 1975 introduced electronic fuel injection as opposed to their earlier mechanical systems, with Cadillac and Lincoln leading the way. These were crude non-feedback designs that utilized very slow analog microprocessors. The next foray by the industry was to feedback carburetors, with a transition to throttle-body injection, and then on to today's multipoint fuel injection.

What does the future hold for fuel management? The next logical step is the development of direct-port injection, where the fuel is introduced under extremely high pressure directly into the combustion chamber, allowing for mixtures in the 40:1 range.

With computer technology making great strides in the 1980s, it became possible to interface electronics with fuel management, and today's fuel-injection systems were born. Not only did this prove to be a major factor in complying with the EPA’s standards, but it also spawned a musclecar renaissance with the arrival of the 5.0 Mustang, Buick GN, and TPI Chevrolets and Pontiacs. It was just a matter of time before these late-model

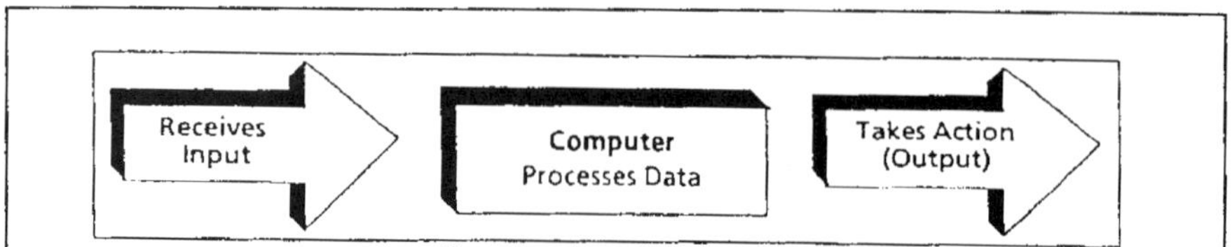

Three Functions of a Computer

A computer processes information and makes a decision based on what it's told.

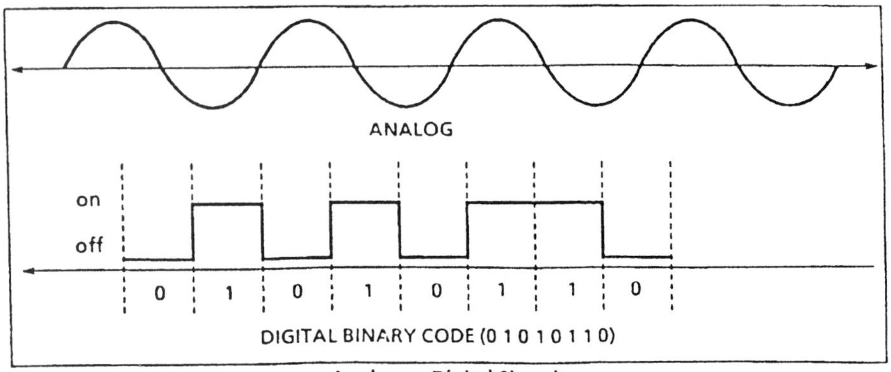

Analog vs Digital Signals

A digital signal creates what's commonly known as a "square wave." The length of time that the wave is either high or low is used by the ECU to identify a parameter. A square wave is what the ECU uses to determine engine rpm when utilizing an electronic spark control ignition. In contrast, an analog signal produces a varying voltage. The TPS and all of the other sensors create analog outputs that are recognized by the ACCEL ECU. Note: The analog signal shown here represents an AC sine wave that experiences a crossing of the polarity line that's referred to as a "zero crossing." Not all analog signals are sine waves but they can remain in either the positive or negative state.

CARBURETOR vs. INJECTION COMPONENTS

Carb	Fuel Injection
A) Float and bowl	A) Fuel rail and pressure assembly regulator
B) Choke circuit	B) Coolant and air charge sensor
C) Fast-idle cam	C) Idle air control (IAC)
D) Idle circuit	D) Base injector pulse width
E) Main metering	E) Base injector pulse width
F) Power enrichment	F) Manifold absolute pressure (MAP) sensor
G) Accelerator pump	G) TPS

injection systems found their way under the hoods of old hot rods, only to be met with, at best, disastrous results. The age of electronic hot-rodding was born, and it was quickly discovered that changes needed to be made to the electronic control unit's (ECU) stock calibration so it would work with traditional speed parts. Due to the complexity of the original equipment (OE) ECU and its proprietary nature, reprogramming by the enthusiast was impractical. That's where ACCEL stepped in and developed a user-friendly reprogrammable ECU, using CalMap software. The deficiencies of air and fuel flow were also addressed with a complete line of manifolds, throttle bodies, and camshafts for small- and big-block Chevrolets, along with fuel injectors and pumps. ACCEL recognized that supplying parts alone would not be enough, so the Engine Management Installation Center (EMIC) program was established as an outlet to support the new age of microprocessor hot-rodding.

Since an engine has no idea how it is being fueled, most of the dynamics remain constant, whether it has fuel injection or a carburetor. What's important about electronic fuel injection (EFI) is that heat is no longer needed in the intake manifold to help vaporize cold fuel, and load-based fuel enrichment becomes part of an electronic decision-making process in lieu of a dynamic action.

The best EFI tuners understand carburetors. Comparing the circuits of a carburetor to its EFI peers is the best way to start to understand ACCEL fuel injection. See the chart at left for a comparison.

The ECU

In any electronic fuel-injection system, the ECU is the core component. Since the ACCEL fuel-injection system uses readily available OE hard parts, only the ECU is unique. It is a Motorola-based microprocessor that employs a Packard head connector along with

The Super Ram intake manifold is an emissions-legal conversion for both GM L98 and LT1 engines

internal components that are reliability rated to 70 degrees C. It was designed and manufactured by ACCEL, with final assembly taking place in an Echlin-owned and operated facility in Independence, Kansas. The ECU is available in two distinct forms; the difference is the injector firing pattern. The most popular unit is referred to as a spark/fuel or batch-fire ECU and is designed to pulse all of the injectors at once. The companion ECU is designated SEFI, which is an acronym for sequential fuel injection. It fires the injectors in time with the valve events. Both ECUs have part number variations for ignition compatibility and desired options. Either ECU can control fuel and ignition timing, and these capabilities will be discussed in depth later on in this book..

The batch-fire ECU incorporates four internal peak-and-hold 4-amp/1-amp drivers and can run eight low-impedance injectors at full-duty cycle for short periods of time. The SEFI unit is capable of controlling eight low-impedance injectors with no time or duty-cycle constraints. A driver can be thought of as an internal switch or relay that has no moving parts; it is

015013 WIRE LIST

CIRCUIT	GAGE	GOES TO	COLOR	NOTES
P1 – D5	18	TPS–C	YELLOW	TPS +5V R
P1 – C1	18	TPS–B	LT. BLUE	TPS SIGNAL
P1 – D1	18	TPS–A	DK. GREEN	TPS RTN–GRD
P1 – D6	18	MAP–C	GRAY	MAP +5VR
P1 – C3	18	MAP–B	ORANGE	MAP SIGNAL
P1 – D3	18	MAP–A	LT. GREEN	MAP RTN–GRD
P1 – D15	16	INJ–A	BROWN	INJ BANK A–H
P1 – D16	16	INJ–B	ORANGE	INJ BANK B–H
P1 – C15	16	INJ–A	TAN	INJ BANK A–L
P1 – C16	16	INJ–B	YELLOW	INJ BANK B–L
	16	GND–T1	BLACK	INJ RTN
P1 – D14	16	O1–B	BLACK	RTN
P1 – C14	16	GND–T1	BLACK	INJ RTN
P1 – C8	16	PWR	RED	+12V BAT
P1 – D7	18	PWR–T2	PINK	+12V SWITCHED
P1 – C12	18	IAC–A	YELLOW/RED	A HIGH
P1 – D13	18	IAC–C	YELLOW/BLACK	B HIGH
P1 – D12	18	IAC–B	BLUE/RED	A LOW
P1 – C13	18	IAC–D	BLUE/BLACK	B LOW
	18	H20–A	GREEN/YELLOW	H2O RTN
P1 – D2	18	AIR–B	GREEN/YELLOW	AIR RTN
P1 – C2	18	H2O–B	PURPLE	H20 SIGNAL
P1 – C11	18	IGN	WHITE	IGN PICK–UP
P1 – C7	16	GND–T2	PURPLE/BLACK	PWR RTN
P1 – C6	18	HUD–F	RED	POT +5VR
P1 – D4	18	HUD–E	PURPLE/WHITE	POT RTN
P1 – C4	18	HUD–D	YELLOW/BLUE	POT 1 SIGNAL
P1 – D8	18	HUD–A	GREEN/ORANGE	POT 2 SIGNAL
P1 – D11	18	RLY–A	WHITE/BLACK	PMP RELAY
	16	PUMP POS	RED/WHITE	PUMP POS
R1 – 3	16	O2–A	RED/WHITE	+12V SWITCHED
	16	RELAY	RED	+12V BAT
R1 – 5	16	PWR	RED	+12V BAT
R1 – 2	18	RELAY,GND	BLACK/WHITE	RELAY, GND
P1 – D9	18	STARTER	BLUE	CRANK DISCRETE
P1 – C5	16	O2–C	RED	SIGNAL
P1 – D10	18	AIR–A	PURPLE	AIR SIGNAL
P2 – A11	18	CALMAP–A	ORANGE	ECU RxD
P2 – A10	18	CALMAP–B	GREEN	ECU TxD
P2 – A8	18	CALMAP–C	RED	ECU +5VR
P2 – B8	18	CALMAP–D	BLACK	ECU RTN
P2 – B2	18	DIST–C		BYPASS
P2 – B3	18	ESC–C		KNOCK CONTROL
P2 – B1	18	DIST–A		EST
P2 – A2	18	DIST–B		REF
P2 – A12	18	DIST–D		GND
P2 – B9	18			NOS ENABLE
P2 – B11	18			NOS SOLENOID
P2 – B5	18			FAN
P2 – B4	18			CONVERTER

This is the pin-out list for a batch-fire ECU.

the component that powers the injector. Since the full current of the circuit being operated flows through the driver, it must be compatible with the fuel injector.

Through the use of ACCEL's CalMap software and an IBM-compatible PC, either ECU is easily reprogrammed and can be configured to work in either open- or closed-loop fuel control with a target air/fuel ratio of stoichiometric. In addition, the ECU can be configured for four, six, or eight cylinders; speed-density-based one-, two-, or three-bar manifold absolute pressure (MAP) input; or a throttle-angle load calculation called Alpha-N.

The ECU can be housed in either a sheetmetal enclosure or an extruded case for packaging in more hostile environments, where water infiltration

With the cover removed from a 6.0 box, the drivers are seen on the upright-mounted heat sink. The four drivers to the left are for the injectors; the other two are for the TCC and coolant fan.

ASAP Inc. in Michigan took over all electronic assembly for ACCEL from the Kansas location. This machine produces plastic header connectors for ACCEL, as well as other customers.

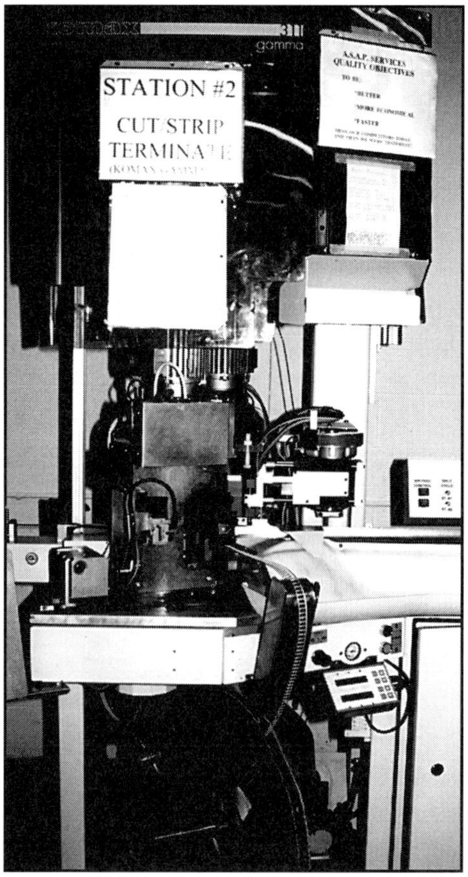

The main wiring harness terminal ends are make by AMP and are cut and stripped at ASAP Inc. to make ECU harnesses.

is an issue. Common with an OE ECU, the DFI uses five-volt analog sensors. Currently, no DFI Version 6.0 ECU offers self-diagnostics.

ACCEL markets its ECU as a stand-alone that can use an existing OE wiring harness with an outside vendor-supplied interface harness, or as a complete kit with ACCEL-produced main and injector harnesses. All ACCEL wiring harnesses are made to meet or exceed OE-quality standards and incorporate Packard connectors.

ECU Inputs

The header connectors are identified by their respective number of cavities: the 32-cavity is identified as P-1, and the 24-cavity is P-2. On each individual connector, the pin locations are identified alphanumerically. For example, connector P1-C8 is 12-volt battery power. The alphanumeric code is molded into the plastic connector and may require some searching to identify the first time. The following is a overview of each sensor and its input to the ECU.

Coolant Temperature—Mounted in the intake manifold water jacket, the calibration is coded to a Delco two-wire sensor. It is considered a thermistor, which is the opposite of a resistor. A resistor's impedance increases when heated, whereas a thermistor's resistance decreases when heated. A five-volt analog signal originating at pin P1-C2 is sent out to the coolant sensor and returns at pin P1-D2. By design, if the circuit fails to open, the ECU will interpret this as extremely cold coolant. In contrast, if the circuit becomes shorted, a coolant temperature of approximately 400 degrees F would be signaled. The coolant sensor value is extremely

The coolant sensor is a thermistor and is sourced from Delco.

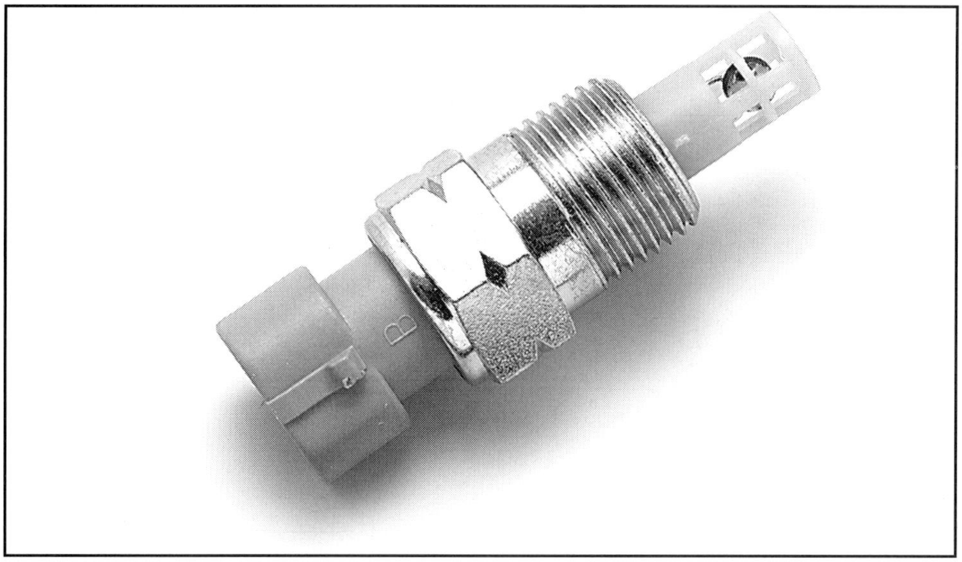

The air charge temperature sensor is also from Delco. It has the sensing bulb exposed for faster response to changes in temperature.

Temperature to Resistance Values for Accel Coolant Sensors

Degree F	Degree C	OHMS
210	100	185
160	70	450
100	38	1800
70	20	3400
40	-4	7500
20	-7	13,500
0	-18	25,000
-10	-40	100,700

Delco MAP sensors are identified by the color of the insert for the wire connector.

important and is used to invoke cranking fuel, warm-up fuel, IAC starting position, fast idle speed, closed-loop threshold, and idle spark functions. For installations on engines other than GM, the DFI-supplied sensor is required.

Air Charge Temperature—Identical in appearance and function to the coolant sensor, the air charge sensor is mounted in the plenum and calculates incoming air temperature to be used as a correction factor. Air temperature correction data can trim the base fuel map to compensate for the decrease in volumetric efficiency (VE) from heating of the charge air. Air is elastic and will contract and expand when cooled and heated, respectively. Hot air has fewer oxygen molecules and would require a smaller amount of fuel to maintain the same air/fuel ratio. Five volts analog originate at P1-D10 and share return P1-D2 with the coolant sensor.

Manifold Absolute Pressure (MAP)—Sourced from Delco, this sensor is available in one-, two-, and three-bar configurations and is used as an input to determine load on the engine. It is connected to the ECU with three wires and is analog in operation. The output of the sensor signals change in the intake manifold pressure. Since the input voltage remains at five volts for each version, resolution is cut down proportionally when a two- or three-bar sensor is used. Location P1-D6 supplies the five-volt reference signal. P1-C3 is the output of the sensor, identified as the signal return, and P1-D3 is an internal ground supplied by the ECU. MAP sensor input is used by the ECU to calculate engine load in both the

A slotted TPS is used so that it can be adjusted to 0.5 percent at closed throttle.

Installation of Lambda oxygen sensor in intake manifold (schematic)
1 Sensor ceramic, 2 Electrodes, 3 Contact, 4 Housing contact, 5 Exhaust pipe, 6 Ceramic protective layer (porous), 7 Exhaust gas, 8 Air.

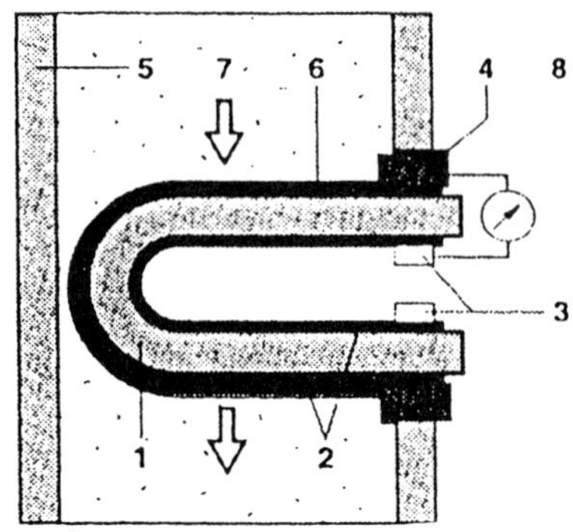

The Lambda or oxygen sensor resides in the exhaust manifold or header and samples the oxygen content to determine the mixture ratio.

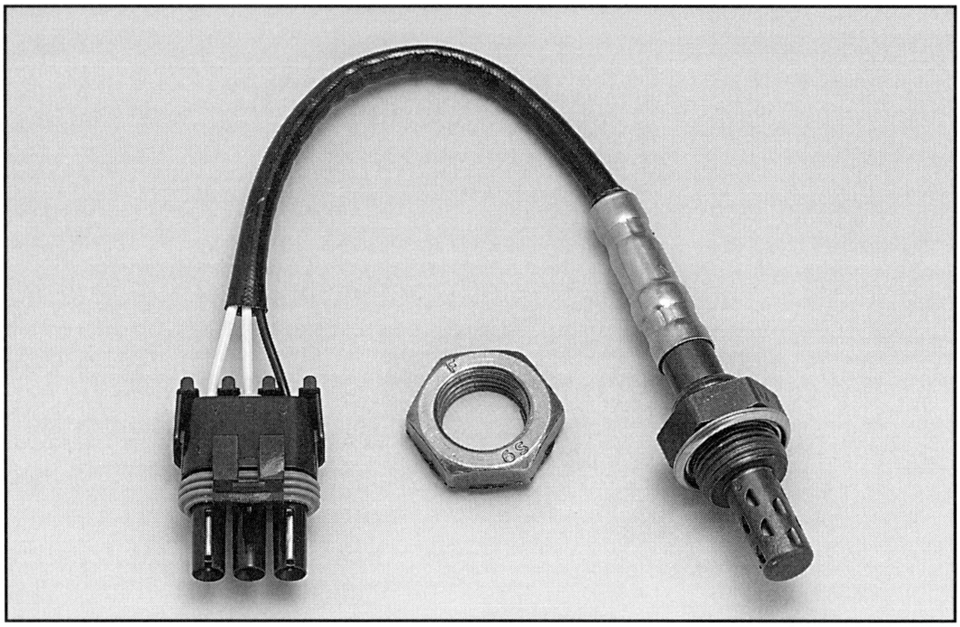
A heated 3-wire oxygen sensor comes with all closed-loop systems.

fuel and timing map, and is also used as an input to invoke load-based fuel enrichment. If the ECU is configured in Alpha-N, then the MAP sensor is retained but is not connected to a vacuum source and is used for barometric pressure compensation.

Throttle Position Sensor (TPS)—Hooked mechanically to the throttle body, the Chevrolet kit uses a three-wire Delco-sourced sensor. Ford kits use a Motorcraft sensor. Regardless of the design, the throttle angle position data is interpreted by the ECU. Circuit P1-D5 supplies the TPS with an analog five volts, while P1-C1 is the sensor's output to the ECU. P1-D1 is an internally supplied ground. Throttle position input is used to promote IAC (Idle Air Control) tracking and also to trigger acceleration enrichment, taking the place of an accelerator pump. Not only is the position of the throttle monitored, but so is the rate of change to determine the amount of enrichment needed. Its input is also used as a threshold for the transition out of closed loop, to engage and disengage the torque converter clutch, and to invoke the idle spark function. Additionally, at 99 percent TPS, the coolant fan circuit controlled by the ECU shuts off, if engaged. During Alpha-N configuration, TPS input replaces MAP to determine load for fueling and spark commands. A slotted TPS is required so base values can be adjusted to correlate with closed and wide-open throttle. Ford installations retain a Motorcraft-sourced sensor, with the mounting holes elongated to make them adjustable.

Oxygen (0_2) Sensor—Used only for closed-loop stoichiometric fuel control, this sensor is a three-wire heated Bosch style. By design, an oxygen sensor will not operate accurately until it reaches a temperature of 600 degrees F. The

Voltage characteristic of the Lambda oxygen sensor at a working temperature of 600°C
a) Rich mixture (air deficiency). b) Lean mixture (excess air).

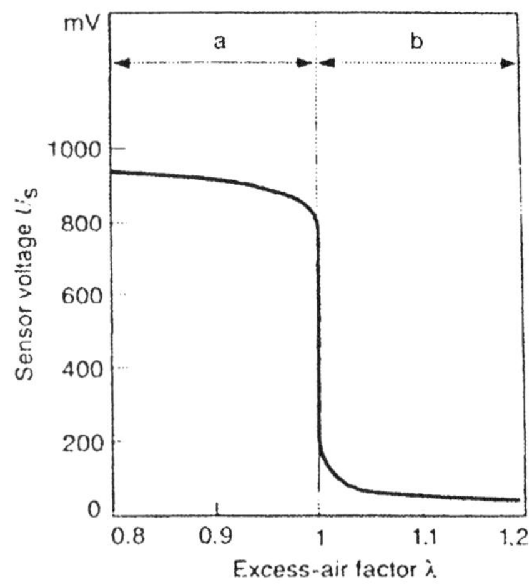

A standard oxygen sensor is only able to accurately identify a mixture of 14.7:1. Note how the voltage hardly changes as the mixture becomes richer or leaner. To determine the air/fuel ratio on this chart, multiply the values on the horizontal axis of the graph by 14.7

Knock-sensor signals
The knock sensor supplies a signal (c) corresponding to the pressure pattern (a) in the cylinder. The filtered pressure signal is shown in (b).

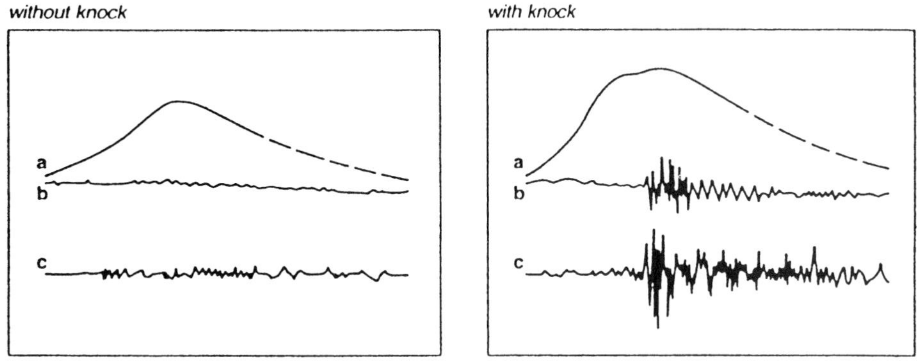

A knock sensor signal during the pressure rise of normal combustion, or when abnormal combustion (detonation) is present.

heating element brings the sensor online sooner and keeps its output voltage more stable. The heater also allows more freedom in the placement of the sensor, which is critical in a performance application that uses headers. Terminal P1-C5 accepts the sensor's output voltage, and the ECU correlates that to an air/fuel ratio. The sensor has an output range from zero to one volt, with .500 being approximately 14.7:1 air/fuel ratio. When the output voltage degrades below .500, the air/fuel ratio is considered leaner than 14.7:1. Conversely, an output of greater than .500 volt represents a mixture richer than 14.7:1. Twelve volts are supplied for the heater from relay terminal R1-3 and are shared with the fuel pump. P1-D4 is the heating element ground supplied internally by the ECU.

***Knock Sensor*—**Optional knock-sensor control kits are available from ACCEL for both small- and big-block Chevrolet engines. The Delco-supplied kit contains a knock sensor, a wiring harness, and an electronic spark control module (ESC), which is essentially an analog-to-digital converter. The sensor is a piezo-electric accelerometer and is calibrated to either engine. Even though they appear to be the same physically, separate part numbers exist for the small- and big-blocks. During detonation, the sensor produces a low-voltage sine-wave output that is converted by the ESC module to a digital signal. The output signal of the ESC module is sent to ECU terminal P2-B3 and is processed internally to retard the ignition timing. (See ECU outputs for further explanation.) Buick GN V-6 applications must source the proper ESC module and sensor directly from Buick. They are not interchangeable with the Chevrolet parts even though they share the same enclosure. The output of the ESC module inputs the ECU at the same terminal for both Buick and Chevrolet. There are no known knock-sensor kits that interface with Ford engines at this time.

***Tach Reference*—**When using an electronic spark timing (EST) distributor, a square wave that originates at the ignition module is sent to terminal P2-A2 to be interpreted as an rpm signal.

***12-Volt Inputs*—**Though not a sensor input, a constant 12-volt source is applied to terminal P1-C8, and a switched 12 volts at P1-D7.

***CalMap*—**ECU terminals P2-A11, P2-A10, P2-A8, and P2-B8 are all used to communicate with the ECU via CalMap software and PC.

By ordering an IPU ECU and an ACCEL 300+ ignition, you can add timing control and EFI to any engine.

ECU Outputs

The following is a list of circuits that are controlled by the ECU.

Injector Firings—A 015013 ECU fires all of the injectors simultaneously, half of their programmed pulse width for each crank rotation. Therefore, this method is identified as simultaneous double fire (SDF). Incorporated into the ECU are four drivers, each with the job of pulsing two injectors on an eight-cylinder engine. The injector wiring harness is segregated into four banks of two injectors each, identified as A-H, A-L, B-H, and B-L. Circuits P1-D15, D16, C15, and C16 supply battery voltage to the injectors; circuits P1-D14 and P1-C14 are returns to the drivers. The drivers supply a ground internally through the ECU.

Fan Control—The ECU has the ability to control an electric cooling fan circuit by grounding an independent relay. The fan-enable temperature is programmable through the CalMap software and operates with a 10-degree F hysteresis. For example, if the fan is programmed to turn on at 180 F degrees, it will not turn off until the coolant reaches 170 F degrees. In addition, as stated in the ECU input section, at 99 percent TPS, the coolant fan is automatically shut off. The fan control is grounded through the ECU at terminal P2-B5.

Fuel Pump—The ACCEL main wiring harness incorporates a Bosch micro-relay to power an electric fuel pump and supply 12 volts to the heated oxygen sensor. Terminal P1-D11 supplies 12 volts to close the fuel pump relay and to turn the pump on. As soon as the ignition is turned on, P1-D11 is supplied with a two-second prime signal to charge the fuel rail. If an rpm signal is not seen on terminal P2-A2, the relay is shut off. Once P2-A2 becomes active again, the fuel pump is turned back on.

Timing Control—The ECU has the ability to interface with either a GM EST module or a Ford EEC-IV ignition for programmable timing control. Timing control can still be obtained on other brands of engines, or if an aftermarket ignition such as the ACCEL 300+ system is used in lieu of an OE ignition module. When terminal P2-A2 experiences an input that equates to over 400 rpm, it applies five volts to terminal P2-B2. Referred to as the bypass circuit, it takes control of the ignition module. Then P2-B1 controls the timing to the ignition module by varying the frequency of the square-wave output. The circuit is completed with P2-A12 supplying a ground.

Idle Air Control (IAC)—The ECU has the ability to have closed-loop idle speed control through the use of a Delco stepper motor air bypass valve. This is a four-wire design that uses pins P1-C12, D13, D12, and C13 to alternate the pulsing of an internal ground circuit to command the idle air control position. The position of the motor is registered in counts, but the ECU is not actually aware of the IAC pintle position; it acknowledges how many commands it has issued from a zero starting point. Installations other than GM require the use of a Delco IAC if idle speed control is desired. ACCEL offers a retrofit kit for Ford applications.

Torque Converter Clutch (TCC)—An optional torque converter clutch engagement kit for GM 200 and 700 series overdrives is available that includes a relay and a wiring harness controlled from port P2-B4. A ground is supplied through the ECU, and the engagement of the TCC is controlled with CalMap software.

Nitrous Control—In its standard form, a 015013 ECU has the ability to

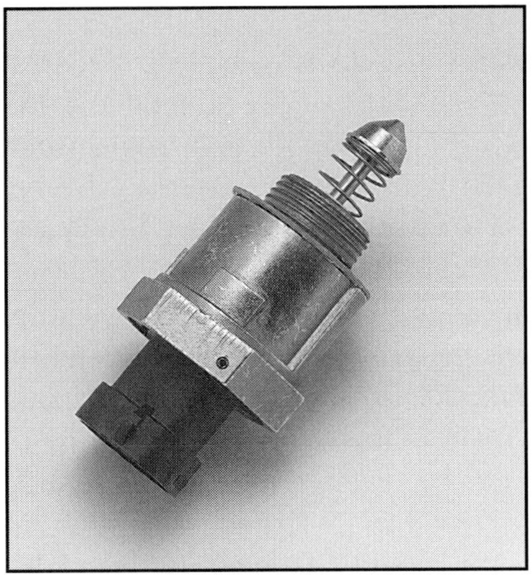

The idle air control motor is from Delco and is a stepper design.

ACCEL EFI has proven that it can take the rigors of all types of racing. This debunks the myth that EFI is not reliable.

control one stage of nitrous. It needs to see 12 volts applied from the vehicle's nitrous-enable switch to terminal P2-B9 to become active. When the parameters are met, which are programmed through the CalMap software, circuit P2-B11 becomes a ground for the nitrous solenoid. Up to three stages of nitrous can be controlled by sacrificing either the TCC or cooling fan control. This modification needs to be done by ACCEL. When the nitrous is engaged, the ECU will automatically retard the timing and richen the fuel curve to the values programmed by the user.

The ECU's Decision-Making Process

Now that you are more familiar with the DFI ECU, let's discuss some of the logic it uses to make control decisions.

The ECU is based loosely on the GM TPI ECM, but it uses a proprietary version of a speed-density-coded software to interpret load and to allow on-the-fly end-user reprogramming. The reason behind the decision to use speed density was rooted in the fact that a mass air flow (MAF) sensor posed three major concerns: the room needed to package the sensor, the flow limitations of the sensor, and an exceptionally high failure rate.

A stock GM TPI MAF is capable of flowing approximately 550–600cfm at 28 inches of H_2O. A fully ported MAF will approach 750cfm. Another obstacle of the MAF is the inability to interface with unique manifold designs. For example, an individual runner intake. Also, the poor-quality German-made Bosch-supplied MAF sensor that GM used on its early TPI systems had a nine-out-of-10 failure rate. By employing a quality American-made Delco MAP sensor, all of these issues were addressed at a substantially lower cost.

The unique aspect of the DFI ECU is not its actual operation, but its ability to allow the end user to tune the fuel and spark curves with relative ease. This procedure is accomplished with the CalMap software and an IBM-compatible PC. Even though the software is menu-driven and extremely user-friendly, the actual tuning decisions require a complete understanding of how an engine and fuel-injection system function. With CalMap you can change the fuel and spark curves along with additional functions of idle speed, warm-up enrichment, air temperature correction, etc. CalMap gives you access to these areas, but it does not tell you the proper numbers to input.

Before we cover programming decisions, let's follow the process the ECU uses to gather data. The goal is to have you thinking like the ECU.

With any electronic fuel-injection system, the ECU gathers information from its sensors and uses it to identify a position in what is known as a look-up table. Think of a look-up table as the chart that is used in a road atlas to determine mileage between two points. For example, if we needed to find the distance between NYC and Los Angeles, a mileage chart would help us figure out this information. These tables, when pertaining to the ECU, are called a matrix, or a map, and exist for both the injector pulse width and the ignition timing control.

As an inanimate electronic device, the ECU understands only voltage and has no way to interpret

CORE TABLES	TRIM TABLES
Cranking fuel	Idle spark
Base fuel map	Coolant enrichment
Timing map	Warm-up enrichment
Idle speed	Coolant temperature vs. idle speed
Configuration	Oxygen sensor correction
Air temperature correction	
MAP and TPS enrichment	

The ECU uses the trim tables in conjunction with the core tables, so it's essential to have the values in the core tables as close to correct as possible because the trim tables have limited control there.

mechanical conditions. For this reason there needs to be a component that can convert a mechanical conditional to an electrical signal for interpretation by the ECU. This is known as a sensor. For example, the ECU does not understand wide-open throttle, but it does recognize a five-volt output on the TPS as this condition.

The ECU has no way of determining if a sensor's output is accurate, so it takes that value as truth. As with any computer, garbage in is garbage out, and proper decisions can only be made with accurate data inputs.

In other terms, what the ECU does is gather data from all of its sensors and then use that data to find a place in the fuel and spark look-up tables to determine an injector pulse width or spark advance command. The DFI ECU requires a MAP sensor input and rpm signal to identify a location in the fuel and timing look-up tables. Other sensor inputs are used as trim tables to the core fuel program. For example, if the look-up table identified a load and RPM cell that placed the base injector pulse width at 10 milliseconds (ms), and the coolant look-up table identified at that coolant temperature should be an additional 10 percent enrichment, the gross injector pulse width would be 11 ms. The base pulse plus or minus the trim value is equal to the gross pulse width. What the CalMap software lets you do is change all the values in the look-up tables to suit the demands of the particular engine.

Now the tables in the ECU can be identified and placed in two categories: core and trim tables.

Air/Fuel Ratio

It's the fuel injector's job to atomize the fuel because liquid gasoline will not burn. For ignition to take place, the fuel needs to be atomized, emulsified, and vaporized, which can be described as broken into small particles, mixed with air, and transformed into a gas. Not to be confused with gasoline, a gas is matter in its most rarefied state.

For an engine to produce power, it needs to burn fuel, but that is in direct contradiction to the phrase "lean is mean." How can we make more power with less fuel? The job of the fuel injector, in conjunction with the ECU, is to meet the air/fuel ratio requirements of the engine. This is an especially challenging task since an automobile engine is required to work at various loads, throttle angles, and rpm conditions.

The term air/fuel ratio is used very liberally when discussing EFI and needs to be fully understood before we can proceed. In simple terms, it describes the amount of air that is mixed with a constant part of fuel. Let's begin with the *stoichiometric ratio* of 14.7:1. That means 14.7 parts of air should be mixed with one part fuel. This shows that as the amount of air goes numerically lower, the mixture is biased toward fuel (richer) since the constant is diluted less. Likewise, if the ratio goes numerically higher, the mixture is being excessively diluted (leaner).

Defining Stoichiometric Ratio—What's interesting is that an engine requires a different air/fuel ratio for various load conditions. What's best for fuel economy is not the ratio for maximum performance. Let's begin by explaining the term "stoichiometric." Abbreviated as stoich, it defines the ideal ratio for the optimum release of chemical energy from the fuel. Often this is simplified by stating that it is the ratio for the best burn, which in reality has removed a little too much from the definition. Unleaded gasoline has a stoichiometric value of 14.67:1 and is commonly rounded to 14.7:1. The ratio is different for other fuels: nitromethane is 1.7:1, methanol is 6.45:1, ethanol is 9.0:1, and 100-octane race fuel is 15.1:1.

In most engines, at peak torque the required ratio for gasoline is around 12.7 to 12.9:1, and at maximum horsepower it's 12.9 to 13.2:1. That is just around 13 parts of air to one part of fuel. During full power, the engine

Many hours of bench testing have gone into the development of the 6.0 ECU.

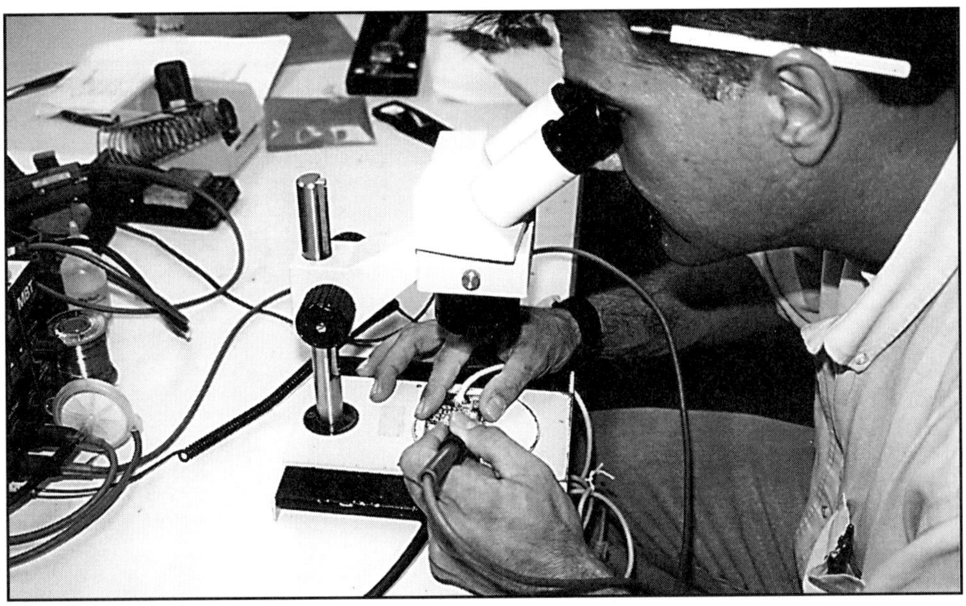

A worker at ASAP Inc. solders a connection on a CalMap circuit board using a magnifying scope for exacting quality. Much of the assembly work on the DFI is done by hand by skilled technicians.

will induct a constant amount of air, which is limited directly by the piston displacement.

If the fuel flow is increased by changes in injector opening or size, the power will rise due to the increased liberation of the chemical energy. The output of the engine will continue upward with an increase in fuel until all of the oxygen in the cylinder is effectively utilized for the combustion event. This will represent the optimum release in chemical energy when ignited. Since the fuel flow can be increased, but the airflow of an engine is fixed by design and displacement, it is the air, and not the fuel, that imposes limits on power.

Maximum power is obtained when there is just enough fuel to consume all of the air in the cylinder. It must then be recognized that the fuel and air are never perfectly mixed with any EFI system, that the fuel may not be completely vaporized, and that some of it is falling into crevice regions. Additionally, the cylinder almost always has a certain faction of exhaust gas present that effectively dilutes the concentration of fresh charge. Because of this, a mixture ratio substantially richer than stoichiometric is required to make power.

At Idle

An especially interesting phenomenon with regard to the required air/fuel ratio occurs at idle. An idling engine demands a mixture that, in relative terms, is richer than at part-throttle. As the throttle plates are opened, the mixture can be made leaner. An explanation for this can be found by noting the pressure in the intake manifold and cylinder bore while idling. When the throttle plate is nearly closed, the pressure in the intake manifold is usually far below atmospheric, whereas the pressure at the end of the exhaust stroke is close to atmospheric. Just as the intake valve opens, there is a higher pressure in the cylinder than in the intake manifold. This causes the relatively high-pressure exhaust gas to expand into the intake tract. As the piston starts to descend on the induction stroke, the reverted exhaust gas is drawn back into the cylinder along with the fresh charge. Thus, the overall mixture contains a high percentage of exhaust gas.

To offset the effects of the inert exhaust gas diluting and weakening the burnable charge, the mixture

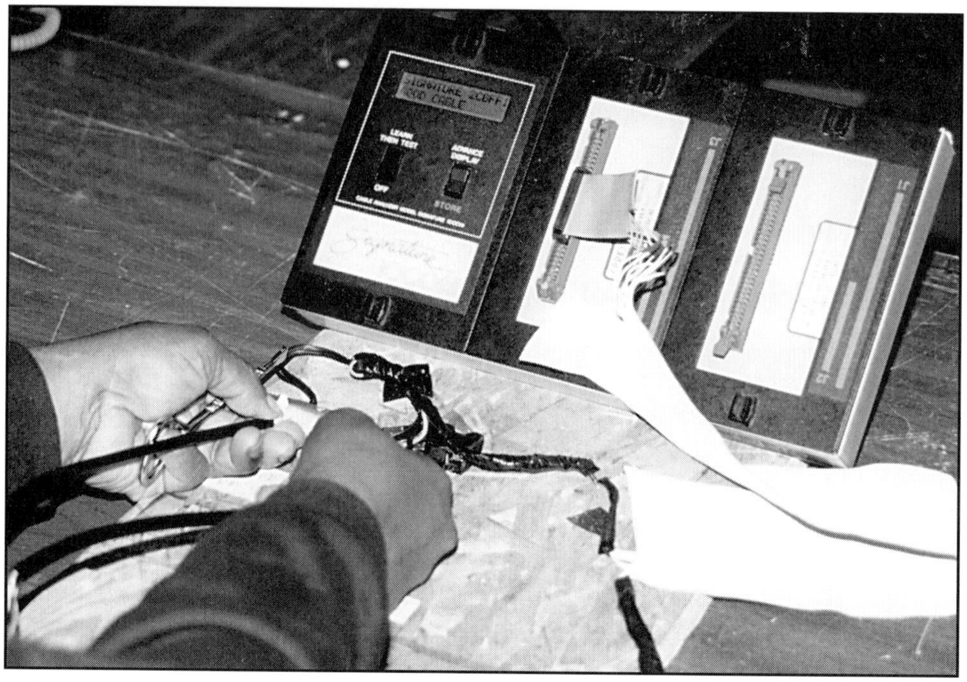

Each electrical component is checked for quality control. This ensures the product you buy from ACCEL will work correctly when installed on your car.

created at the fuel injector needs to be rich. Though it has been well established that the pressure and temperature of the exhaust gas at the end of the exhaust stroke do not vary greatly with engine load, the clearance space in the cylinder remains constant, so the amount of trapped exhaust gas stays constant. The concern is the most prevalent at idle, since the throttle plates are almost closed and the engine requires a very small amount of fresh charge by weight to overcome internal friction. The volumetric efficiency (VE) of the engine is at its poorest, but the volume of reverted exhaust to fresh charge is high. This condition is present in all engines, even those with very mild cam profiles, but dilution is accentuated when cam overlap is increased. As a slower idle speed is demanded, a richer air/fuel ratio is required, or the idle speed must be raised to allow for an increase in fresh charge.

Defining Vacuum

The movement of the piston creates a low pressure in the cylinder bore and on the back side of the throttle plates in the intake manifold.

The standard used to identify pressure is relative to the force of the atmosphere on Earth, better known as atmospheric pressure. By definition, a vacuum is any pressure less than atmospheric. This is measured in many ways and can often lead to confusion: It can be represented in pounds per square inch, inches of mercury, inches of water, or bar. According to McGraw-Hill's *Dictionary of Engineering*, atmospheric pressure is measured at sea level and is considered to be 14.70 psi, which is also called one atmosphere.

A typical vacuum test gauge reads in inches of mercury. If a vacuum gauge is attached to an engine and reads 15 inches of mercury, this means that the low-pressure region created in the engine is sufficient for atmospheric pressure to push a test column of mercury 15 inches up into a tube. Expanding on this, atmospheric pressure of 14.70 psi can push the same column of mercury 29.92 inches. In near terms, due to differences in specific gravity, a manometer would read 406.78 inches of water at atmospheric pressure. So as you can see, what we accept as either pressure or vacuum is actually relative to the atmosphere.

MAP Sensor**—**The MAP sensor used with the DFI system is calibrated in the metric equivalent of pressure, which is bar. This is defined as a unit of pressure equal to 10 Pascals, or 10 newtons per square meter, or 10 dynes per square centimeter. The MAP sensor reads the difference between atmospheric pressure and the conditions in the intake manifold. A one-bar MAP sensor reads from 15 inches of vacuum to atmospheric pressure; a two-bar MAP sensor will read from 15 inches of vacuum to 14.7 psi above atmospheric; and a three-bar sensor is accurate from 15 inches of vacuum to 29.4 psi above atmospheric. Forced-induction engines need to use a two- or three-bar MAP sensor that can recognize boost, which is any pressure above atmospheric.

2

Fuel Flow Dynamics

All modern electronic fuel-injection systems use an electric fuel pump to maintain the constant supplies of fuel and pressure required for operation. ACCEL offers two fuel pumps for EFI applications: part number 74701 supports engines up to 450 hp; part number 74702 will supply power levels to 800 hp. Extremely high-horsepower engines would require two pumps installed in tandem. Both of ACCEL's pumps can be mounted either in the fuel tank or cell, or externally on the vehicle's frame.

Fuel Pressure vs. Fuel Flow

A common misunderstanding with electronic fuel injection is the concept of pressure versus flow. This relationship is often referred to as a seesaw: When one goes up, the other goes down. Fuel volume is required to maintain pressure. If the fuel volume ramps down under engine load, the fuel pressure will also drop. To understand this better, think of a simple garden hose. In a residential water system, the amount of water volume to the hose spigot remains constant. When a garden hose with a nozzle is attached, it controls the volume along with the discharge pressure of the water with a restrictor. If another load is attached to the same water line, such as an additional hose, what would happen to the flow from the first nozzle? As we all know, the pressure and flow at the original nozzle will decrease. In essence, the same pressure/flow relationship happens with a fuel injector. As load is applied to the engine, an increase in the injector opening time is required to create the proper air/fuel ratio. If the fuel flow is not sufficient, the pressure at the injector will drop. For this reason, fuel pressure should always be checked at idle and then under load to determine if the fuel-delivery system can supply the needs of the engine.

Pressure Regulator

The function of the fuel regulator is to maintain a predetermined pressure in the fuel rail. All electronic fuel-injection systems pump more fuel than the engine consumes, constantly returning the fuel to the supply tank. The fuel pressure regulator is

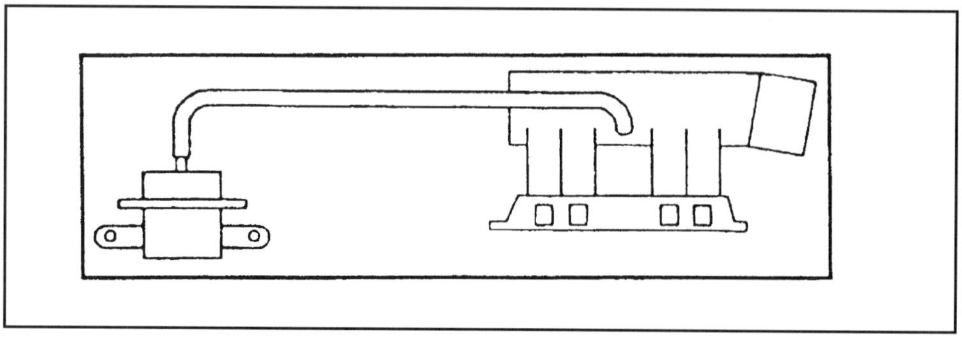

The fuel pressure regulator needs to be referenced to a manifold vacuum source.

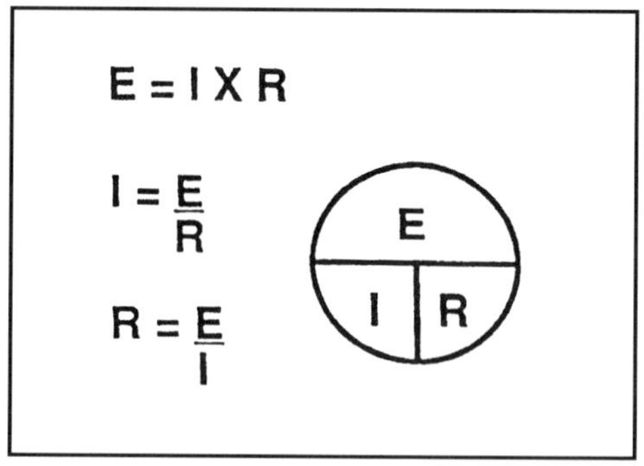

Ohm's law applies to all electrical devices.

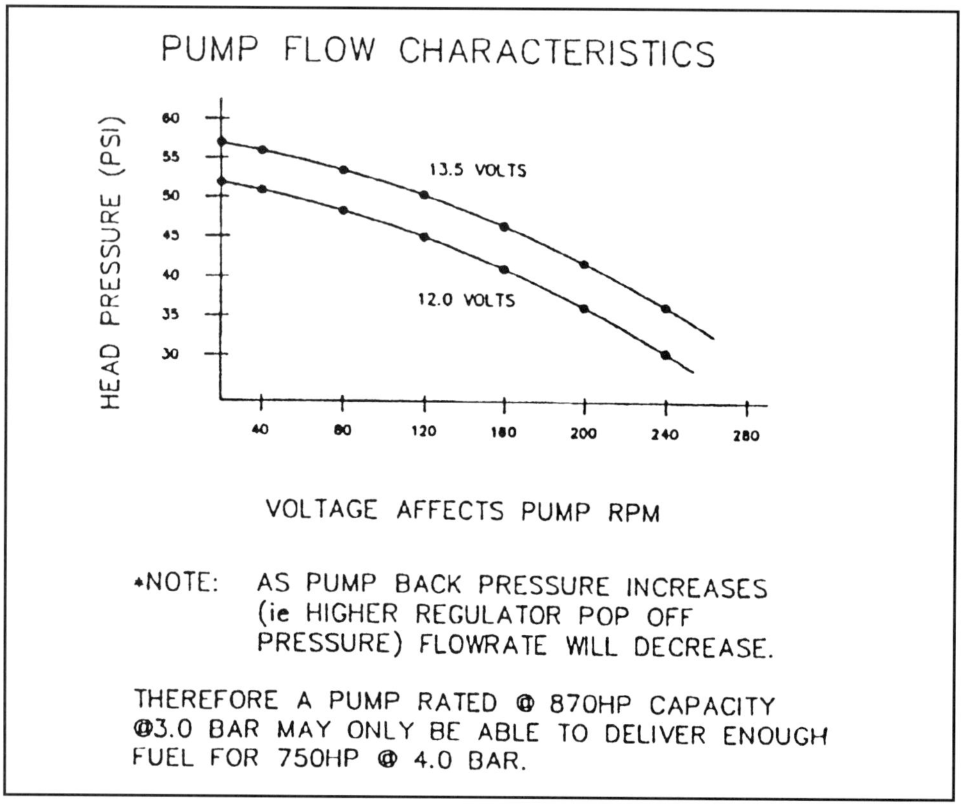

Obeying Ohm's law, an electric fuel pump's output is affected by changes in voltage.

The design of the fuel filter allows for the constant flow of gasoline.

connected in series with the fuel supply and maintains a set pressure by controlling the amount of returned fuel; the less fuel returned, the higher the rail pressure. Referenced to manifold vacuum, the pressure is reduced under high engine vacuum conditions such as coast-down. As the engine vacuum drops, the fuel pressure is increased. Forced-induction applications should have a fuel-pressure regulator and pump with the ability to increase the fuel pressure one psi for each psi of boost. The regulator consists of a vacuum diaphragm, calibrated spring, and vacuum port. An adjustable regulator has the ability to change the diaphragm position and, thus, control the amount of return fuel. If the fuel flow volume is insufficient, then regardless of the diaphragm's position in the regulator, the fuel pressure will not be maintained. Remember, to maintain the fuel pressure set point, there needs to be more fuel supplied than consumed.

Voltage Supply

Another area of concern that affects the fuel-pump delivery is the voltage supply. Most electric EFI pumps are a rotary-vane design that requires a constant supply of voltage to maintain its rated output. Applying Ohm's Law, if the current demand increases, then the voltage will drop if the voltage supply remains constant. The chart at left shows the direct correlation of fuel pump output to supply voltage.

Fuel Rail Hydraulics

Not only does voltage affect pressure, but the size of the fuel supply and return lines and the area of the fuel filter become paramount. Fuel-injected engines always have larger fuel filters in comparison to

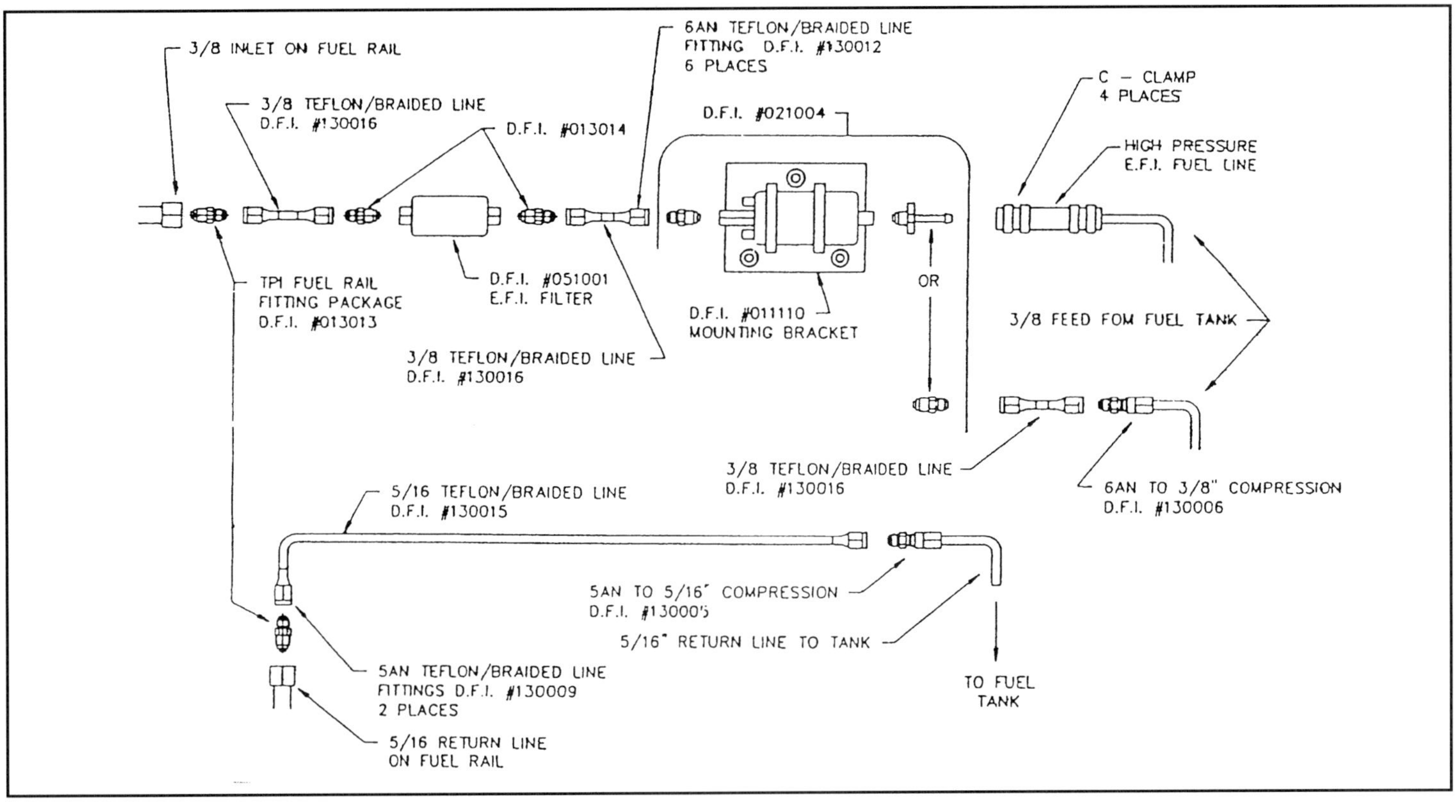

Use this drawing as a guideline for your fuel system installation.

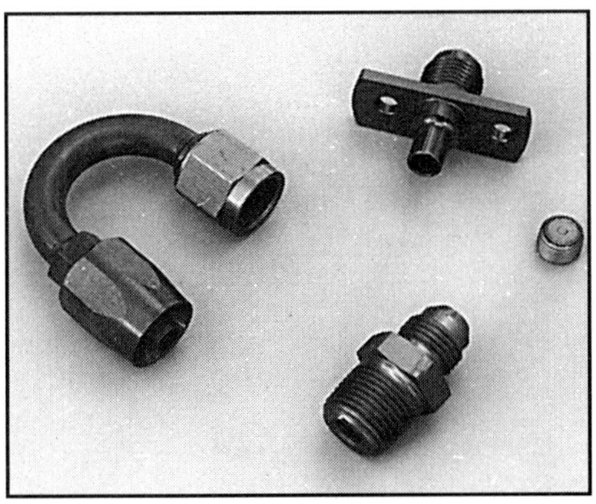

ACCEL Ideal Instant Injection offers this trick fuel rail return kit.

their carbureted counterparts due to the volume of fuel pumped. Remember, there is always more fuel pumped than used, and the filter can not disturb the flow characteristics of the fuel system. Also, keep in mind that a fuel-injection system has no reservoir to store fuel as a carburetor does; the only storage it has is the fuel rail.

The sequence at which the injectors are fired also has an effect on the fuel rail hydraulics. Simultaneous double-fire is the least complicated method electronically, and in theory has less components to possibly fail. But it's the hardest on the fuel-delivery system. When eight injectors open simultaneously, the fuel rail becomes rapidly depleted, causing the fuel-pressure regulator to close to maintain pressure. This creates a hydraulic surging in the rail called *fuel rail knock*, especially when large fuel injectors are used. The knocking sound is produced by the fuel rail being shocked by the rapid discharge of the fuel supply and the rush of replacement fuel. Sequential-injector firings are the most fuel-rail-friendly due to the minimal discharge by opening one injector at a time.

Check Valve—A common feature in an EFI fuel-supply system is a fuel pressure check valve that will maintain fuel pressure to the regulator when the electric pump is shut off. This is used to aid in starting the engine by keeping the fuel lines and rail charged, and to help combat percolation in the fuel rail during heat-soak conditions. ACCEL-supplied fuel pumps incorporate an internal one-way check valve located on the pump's outlet side.

Plumbing and Installing the Fuel System

The best fuel system components in the world are useless if they are installed incorrectly. The amount of attention paid to detail will either make or break your fuel system long after the parts are paid for. The good thing is that a proper installation is not difficult to perform once the theory is understood. There are three distinct areas of concern: component choice, plumbing, and electrical.

If more than one fuel pump is employed, always use a separate relay for each.

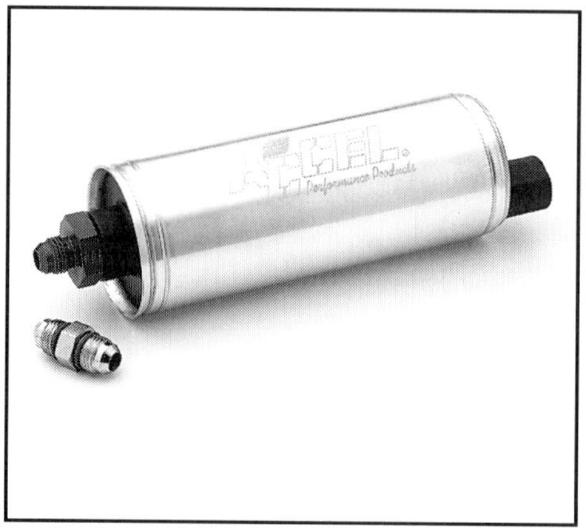

All ACCEL EFI pumps can be either frame- or in-tank-mounted.

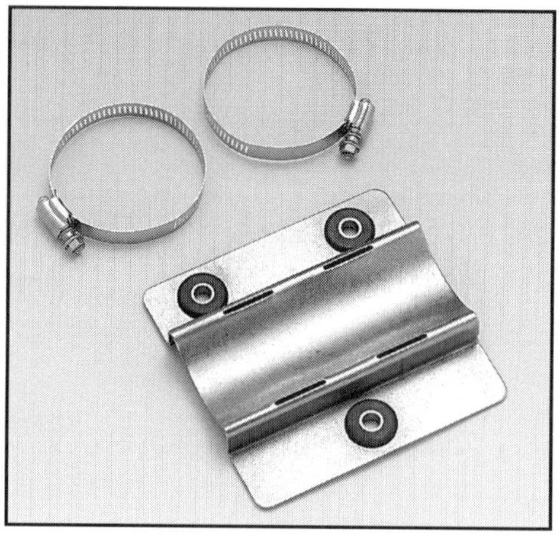

Use part number 74710 for a vibration-free external or frame mounting of a fuel pump.

Fuel System Size

Often a fuel system is sized too large, with the installer working under the theory that the fuel will be there if the engine program is stepped up a few notches. This is not necessarily a faulty logic, within reason, but as with most aspects of this sport, excess seems to be the norm. The problem with using an excessively large fuel pump, especially on a street engine, is the need to control not only the fuel pressure at the rail, but also the return to the tank. If the pump is too large and moves a volume of fuel that is too great for the regulator to bypass, then at light loads the fuel pressure will increase substantially. This is especially common with turbocharged engines that produce great levels of power under boost but are small in displacement and docile when driven normally. These applications have a "Dr. Jekyll and Mr. Hyde" fuel-delivery requirement, demanding a high volume under boost but a minimal delivery rate under cruise. If the dimension of the bypass port in the fuel-pressure regulator and the size of the return line are not considered, the fuel pressure can actually double under light load. Not only will this create an excessively rich mixture, but it will cause premature pump failure due to the high pressure it must work against.

Fuel Line Size and Material—Line size and material are other areas of confusion. A good rule to follow for sizing is to maintain at least the same outlet dimension of the fuel pump. Anything larger on the supply side will not hurt and will add volume through added storage capacity. However, the feed from the fuel cell or gas tank should be at least one complete size larger than the outlet dimension of the fuel pump. Braided steel line may look trick but is usually a very expensive overkill for most installations. A good compromise is high-grade aluminum tubing; it's strong, light, and easier to form than plain steel tubing, though it's more expensive. Rubber line should be limited to the flex area, where the fuel line leaves the frame of the vehicle and is moving to the engine. Many sanctioning bodies limit the amount of rubber line that can be used. For example, the NHRA will not allow more than one foot total of rubber fuel line in the entire vehicle. It is also imperative that the line, steel, aluminum, or rubber is rated for

A typical in-tank fuel pump installation.

ACCEL PERFORMANCE FUEL INJECTOR PROGRAM
Technical Data and Specifications

ACCEL Individual Injector Catalog p/n	Catalog lbs/hr Flow Rating or 39.7 PSI	Static lbs/hr Flow Rating @3.0 BAR or 44.1 PSI	Static cc/min Flow Rating @3.0 BAR or 44.1 PSI	High Impedance Resistance Ohms
150114	14	14.1	135	14.4
150115	15	15.4	150	14.4
150117	17	17.3	170	14.4
150119	19	19.2	185	14.4
150121	21	21.1	205	14.4
150123	23	23.0	225	14.4
150124	24	24.3	235	14.4
150126	26	25.6	250	14.4
150130	30	29.4	285	14.4
150132	32	32.0	310	14.4
150136	36	37.1	360	14.4
150140	40	39.7	385	12.0
150148	48	47.8	465	12.0

gasoline and high pressures. All EFI fuel hose should be rated to withstand at least 150 psi or greater. The return line should be one size smaller than the line from the outlet of the fuel pump to the fuel rail.

Whenever a fluid is asked to make a turn, frictional flow losses occur, limiting flow in gallons per hour while raising the pressure the pump must work against. Although it would be impossible to create a fuel system that makes no turns, some time spent studying and planning your installation can often eliminate many unnecessary maneuvers.

The ideal path would be as straight as possible from the fuel tank or cell to the inlet of the fuel rail. When the direction of the fuel line must change, the gentler the bend, the happier the fuel will be. This means that a 45-degree turn is always better than a 90-degree. If the installation demands a 180-degree turn, it is best to go up one tubing size for that fitting and break it into multiple gentler bends. Often it looks as if the installer thought he was making a moonshine still instead of a fuel system.

Just remember: As the area of the radius increases, frictional flow loss decreases. The style of fitting and connections will also impact flow losses. AN connections are the least restrictive and should be favored over the less expensive NPT style.

When installing the electric fuel pump, always use a relay—either the ACCEL-supplied one in the main wiring harness or one of your own choice—along with a fusible connection. As mentioned previously, Ohm's Law will impact the output of the fuel pump. Since an OE charging system is designed to put out approximately 14.5 volts DC, most fuel pumps are intended to work on at least 13.6 volts DC.

By incorporating a relay from a switched ignition source, the fuel pump can be powered directly from the alternator or battery with proper-size wire. All electrical connections should be soldered and shrink-wrapped for protection against moisture and corrosion. The ground wire of the fuel pump should run back to the engine block at the same point that the battery ground strap attaches. This will ensure that the circuit will operate on the same ground plane with the least resistance.

The Fuel Injector

Other than the ECU, the main component of an injection system is its namesake, the fuel injector. A fuel injector is identified by a number of characteristics: flow capacity, fuel feed point, attachment to the fuel rail, electrical resistance, and tip design. In some applications, another criterion is the placement of the injector. Some manufacturers, such as General Motors, use application-specific designed injectors in different EFI systems. GM port, throttle body, and

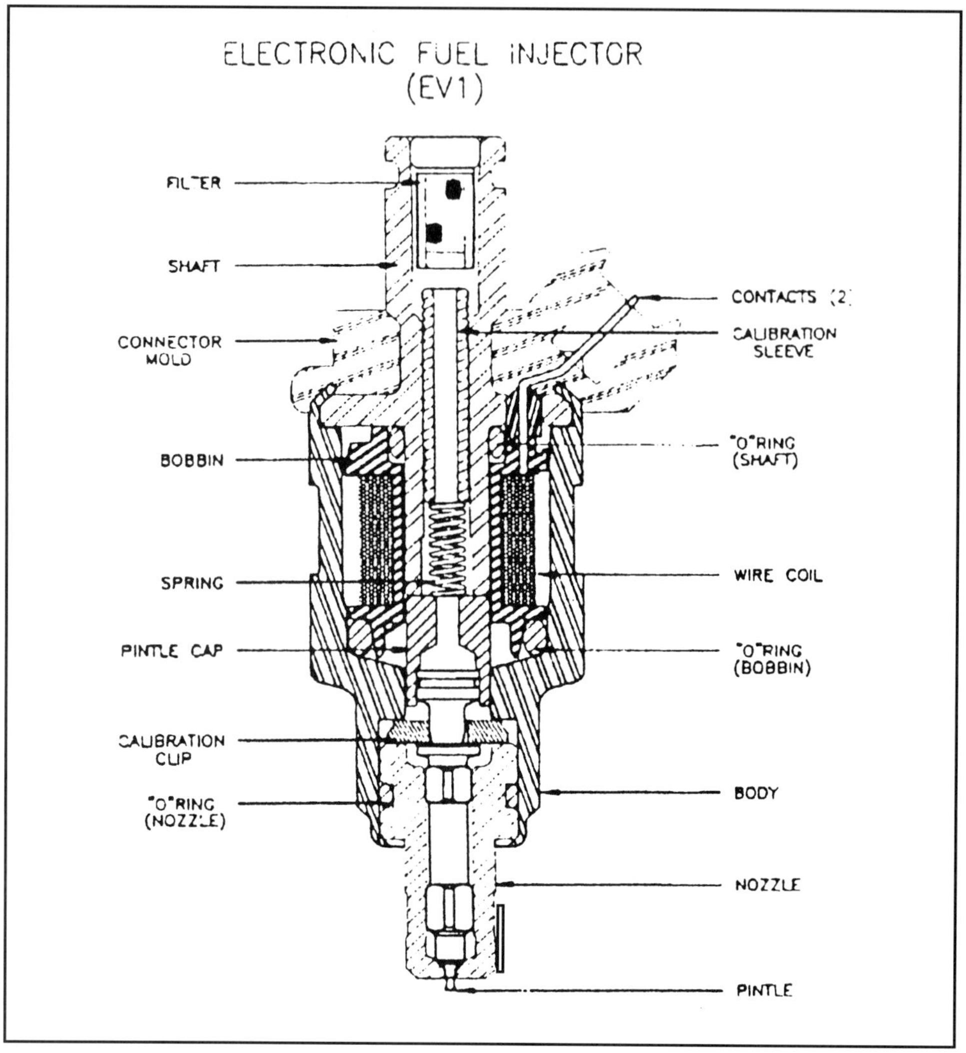

Most modern fuel injectors are patterned from the popular BOSCH EV1 design.

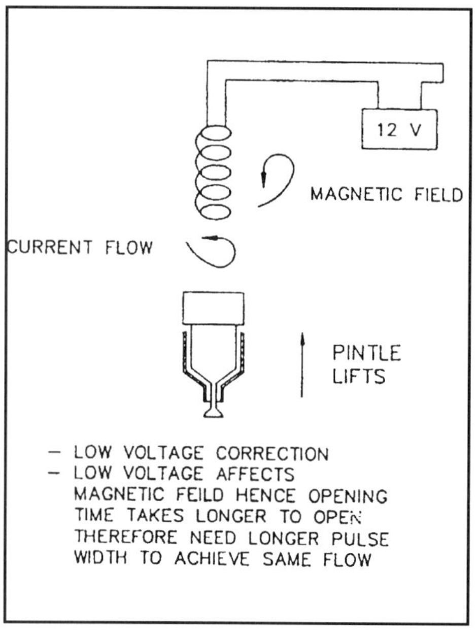

The workings of an EFI injector.

central port-injection all use a unique injector. ACCEL sells only individual port-style injectors.

Before we expound the differences in port-style injectors, ACCEL injectors are manufactured by RAM Pacer and are classified by their flow and resistance values.

All ACCEL injectors are top-feed with O-ring attachment to the fuel rails. Common to all electronic fuel injectors is a solenoid attached to a fixture that opens and closes a fuel-flow orifice. When a ground circuit is completed by the ECU, the solenoid is energized, lifting the fuel-flow device and exposing an orifice to pass fuel. When the ground circuit is removed, the solenoid closes and fuel flow stops. In Chapter 1, the terms "core fuel table" and "programmable amount of injector pulse width" were used. They describe the amount of time the injector circuit is grounded, and that the signal to open the injector was applied. The opening of the injector may seem uncomplicated, but there are other crucial elements that need to be applied.

The length of time the injector ground circuit is applied is measured in milliseconds, or thousandths of a second. This is the injector pulse width. The length of time it takes for the solenoid to lift and completely uncover the fuel-flow orifice is called the rise time. Even though this may seem to happen instantaneously, when measured in thousandths of a second, the slowness of the injector opening becomes apparent.

Ohm's Law states that if the voltage remains constant, an increase in current will charge the solenoid more quickly and make it respond faster. By altering the resistance of the injector windings, additional current can flow and decrease injector rise time.

Most OE applications use high-impedance (12–16 ohms) injectors due to their lower cost and the ability to use inexpensive saturated drivers in the ECU. Lower-impedance injectors (2–4 ohms) respond quicker (shorter rise time) but necessitate the use of peak-and-hold drivers, which are not only more complicated but costly to manufacture. By design, a saturated driver will keep the current draw constant during the cycle. Conversely, a peak-and-hold driver will initially surge the current up and then step it down to a lower value, maintaining that current throughout the event. If a peak-and-hold driver is rated at 4/1

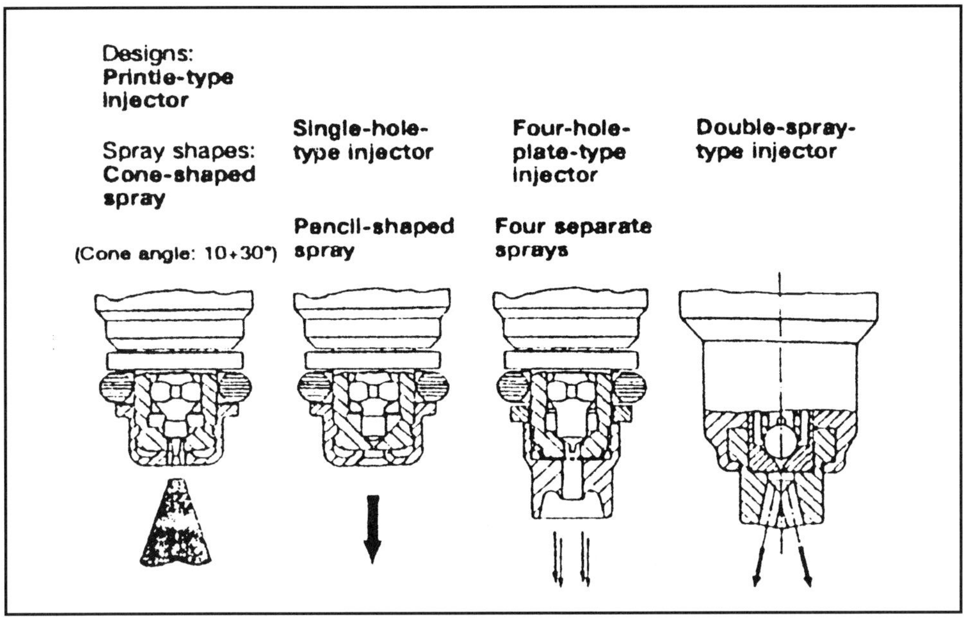

Many different injector spray configurations are used by the OE market. The double-spray styles are for 4-valve cylinder head applications.

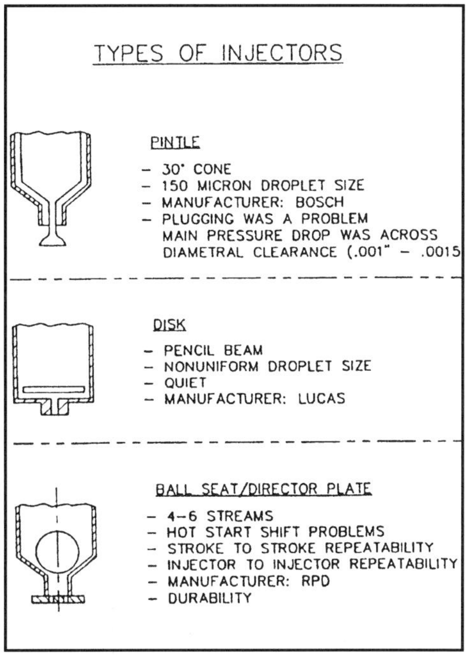

Each injector design offers unique characteristics.

amps, that translates into four amps to open the injector and one amp to keep it opened. Historically, tests have shown that a low-impedance injector will have a rise time of just below 1.5 ms, and a high impedance unit can approach two ms.

Fuel delivery at or below the minimal injector rise time skews the atomization and leads to high emissions and poor idle quality. A standard 015013 ECU has four 4/1 drivers and can operate eight low-impedance injectors for short periods of time. Due to the drivers used in this ECU, long-duty cycles and the fact that it is simultaneous double-fire-coded will cause overheating of the ECU (see the tuning section for further explanation). The ACCEL sequential-fired ECU uses eight individual drivers and has no current or duty-cycle constraints.

In addition, an injector is rated for its flow capacity. Think of injector flow as a carburetor jet orifice dimension. ACCEL uses pounds per hour to rate its injectors. Other standards are grams per minute and cubic centimeters per second. When an injector is rated in pounds per hour, it's always at a test pressure of 43.5 psi, which equates to the metric reading of three-bar. For example, an injector rated at 24 lbs.-hr. would flow 24 pounds of fuel at 43.5 psi pressure. If the pressure is increased above the test value, the injector flow increases. Conversely, if the test pressure degrades, flow suffers accordingly. A good rule pertaining to injector resistance with ACCEL products is that any rating over 36 lbs-hr. would be low-impedance.

The previously mentioned flow rates are not only at three-bar test pressure, but also at 100-percent duty cycle. This brings us to another topic. Duty cycle is not a function of injector design, but it is affected by it. Because an injector has to supply fuel for a multitude of engine operating conditions, it has to be very adaptive. It must respond quickly (rise time) to supply the proper atomization at short pulse widths during idle or light load while not starving the engine at full throttle. This is very difficult to accomplish. The easiest way to understand injector duty cycle is to think of it against the length of ignition event time that the ECU is supplying a ground path for the injector. Since the ECU opens the injectors in time with the ignition events, there is a specified amount of time to allow the fuel into the cylinder before the event for the next cylinder begins. The length of time between ignition events is much shorter at 6000 rpm than it is at 3000 rpm. If the injector is open longer than the ignition event, then it is considered to be static and not able to control fuel flow. That is why injectors are offered in different resistance values and flow rates. If the injector cannot pass enough fuel during one event to satisfy the engine, either the fuel pressure needs to be raised, or the flow capacity of the injector increased. Lower winding resistance allows for quicker response and, in turn, the ability to increase the amount

The engineers at DFI made this test fixture to quantify the spray pattern of different injector designs during development.

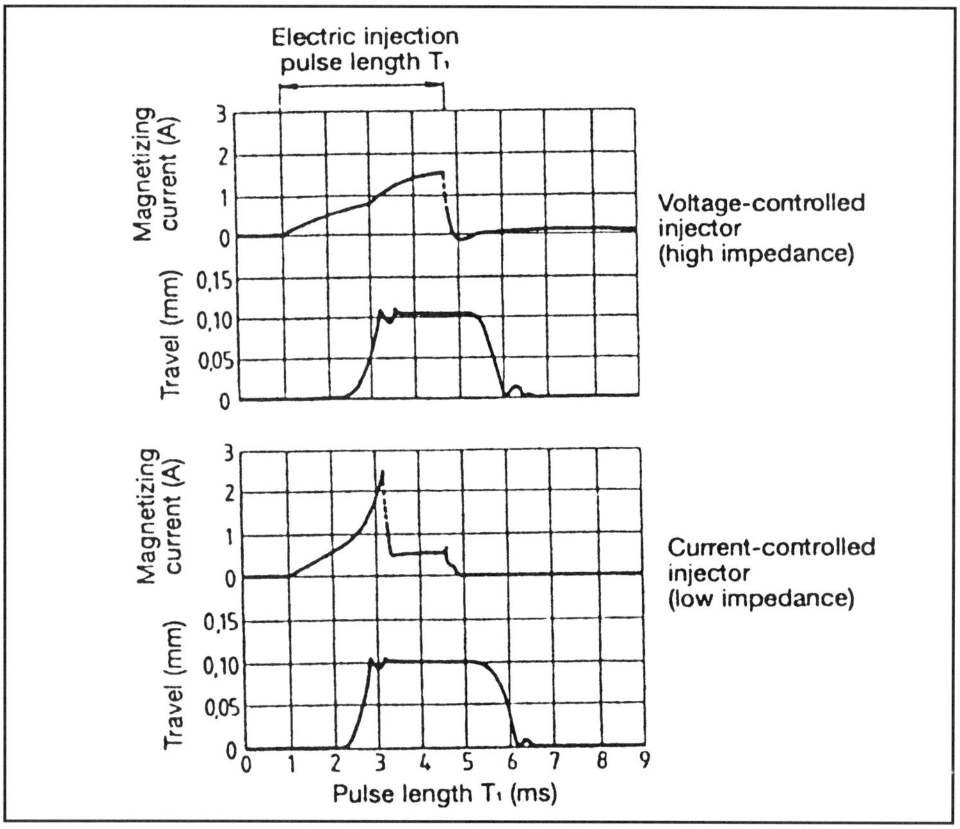

High-impedance injectors take longer to open than low-impedance designs.

of usable time between ignition events. OE logic usually does not use fuel pressure as a tuning aid due to the possibility of fuel leakage past the closed pintle on high-mileage injectors, which would cause excessive coast-down emissions.

In the performance aftermarket,

An ASNU injector flow bench is used at the University of Northwestern Ohio in an EFI class.

injector duty cycle on an engine that will see short bursts of high rpm can run near static and cause no problem as long as the air/fuel ratio can be maintained. When sizing an injector for an endurance engine, accept a maximum of 80 percent duty cycle. This will allow the injector and drivers to cool.

To choose the proper injector size, you need to know not only all of the above, but also the brake-specific fuel consumption data of the engine (BSFC), along with accurate horsepower numbers. This will tell you how much fuel in pounds it takes to produce one horsepower. One rule is that a normally aspirated engine will use one-half pound of fuel (.5 BSFC) for each horsepower produced, whereas a forced-induction engine will consume slightly more fuel at .55 pound. Then again, the equations in this chapter will establish this further, and a more in-depth discussion of factors that affect BSFC will be covered in the tuning section. Using a BSFC value of .5, a 30-lb./hr. injector at three-bar fuel pressure and 100 percent duty cycle can support 60 hp per cylinder, or a total of 480 hp on an eight-cylinder engine.

The design of the injector tip will not affect its flow rate but will impact how the fuel is atomized, especially at short duty cycles.

Rochester-produced injectors use a ball valve with a director plate, and Lucas uses a disk shape. Each injector design has its own weaknesses and strengths, but the pintle design that ACCEL uses has proven to be the most accurate and repeatable out of the three, albeit the most costly to manufacture.

The next area of concern is stroke-to-stroke repeatability and its effect on engine performance. Most people think "out of sight, out of mind" and feel that as long as the injector is pulsing, all is well. When dealing with a port fuel-injected engine, think of it as multiple carburetion.

Anyone who has worked with multiple carbs, specifically of the Weber type on IR runner manifolds, has a full understanding of fuel distribution and the problems it can cause. Since each injector is responsible for the entire fuel requirements for that particular cylinder, any flow variation from injector to injector will cause differences in mixture for each bore. When injectors become dirty, they lose their ability to properly atomize the fuel, causing high emissions and idle instability. With today's fuels, injectors tend to stay cleaner than they did 10 years ago, but tip fouling is still a concern. Even if the tip of the injector does remain clean, the constant charging and discharging of the solenoid will eventually lead to a injector that is lazy electrically, suffering from an increased rise time. The only way to quantify injector performance is to remove the injectors and cycle them on a test bench, such as the Asnu injector tester and cleaner. This will let you view atomization patterns and measure flow accurately. It is not uncommon for out-of-the-box injectors to have flow variations of up to 15 percent, and for this reason ACCEL sells only matched sets of injectors.

Measuring Horsepower and Torque

"How much power does it make?" This question invariably surfaces in every conversation about an engine. Abiding by the laws of physics, power is required to change a body at rest into one in motion, and whatever has the most power goes the fastest. This creates the need to determine an engine's output. Even though power can be calculated mathematically according to a vehicle's rate of acceleration or top speed, the ideal place to define the amount of grunt is at the flywheel or drivewheels during a dyno test. Most dyno sessions are used to sort out and tune combinations or parts; others use the dyno to quantify and qualify the maximum output of the engine. Either way, valid documentation of engine output is the reason for investing in dyno time.

It's important to determine the proper dyno for the job: either engine or drivewheel. The benefit of engine dyno testing is the ease of swapping parts and attaching data acquisition and test equipment that might be difficult or impossible to use in the confines of the engine compartment. A chassis dyno, on the other hand, provides relative ease of testing but limits changes and modifications. Until recently, most chassis systems were limited to wide-open throttle testing and were unable to sustain a given load for part-throttle tuning. Ancillary to this is the ability to record the air/fuel ratio and exhaust gas temperatures. Although working only with power numbers is helpful, it lacks the depth needed to prescribe modifications and tuning help. Ideally, you should define your combination with the engine dyno, then put it in the car and run it on the chassis dyno to do the tune-up.

Power is found in the chemical-to-mechanical energy change that is created from the expansion of the flame front during ignition and through the power stroke. During this time, the cylinder pressure increases as the flame presses down on the piston. The real measure of an engine's ability to do work is the amount of cylinder

continued on next page

pressure produced during the power stroke.

A dynamometer measures torque, which is defined as the amount of work that can be accomplished; horsepower is calculated as work over time. Though torque is a valuable means of measuring the potential to do work, it is affected by the displacement of the engine. For this reason, OE engineers reference mean effective pressure (MEP), a parameter of force-per-unit area, rather than horsepower or torque. This can then be measured in terms of torque, which is referred to as brake mean effective pressure (BMEP). The maximum BMEP of an efficient engine is essentially constant over a wide range of engine displacements. Then the BMEP of an engine can be compared with those of other engines of varied size, and the engine's effectiveness to produce power can be compared with peers of different displacements. The aftermarket does not recognize BMEP but rather compares horsepower to cubic inches (hp/ci).

All dynamometers, except inertia-based chassis types, incorporate a device that applies resistance to the engine as would a brake. It may build resistance via a friction material or electrical current, but most popular ones use hydraulics. Thus, the term "brake" was applied to the readings of both horsepower and torque because it was a measurement of the power consumed by the resistance mechanism. Today, brake has been replaced by flywheel power, delivered power, or drivewheel power. Power may also be registered as net, gross, peak, and continuous.

Before 1971, most OE manufacturers advertised gross power, or what the engine would produce on the dynamometer with intake and exhaust systems that were more efficient than the ones in the vehicle, and without any power-robbing accessories. The net rating, however, accounted for all of the components and accessories from the production car and reflected substantially lower output.

The aftermarket usually publishes gross peak readings because the exhaust and air intake systems from the vehicle are rarely included in the test procedures. Detroit publishes net continuous ratings, which is the power the engine can produce for extended periods without any anticipated failure while using the production ancillaries and drive loads.

Since an engine's ability to do work is affected by atmospheric conditions, correction factors are necessary to compare engine output. A test performed on a low-barometric-pressure day would yield different results than one on a high-pressure day. Likewise, if the air temperature was to vary, the result would be affected, also. Since we do not have the ability to control the atmosphere, the SAE has developed and identified test standards that include correction for temperature, barometer, and humidity. Since most correction factors are set at the dyno console, the phrase "spin the dial until the check will clear" has been applied by unscrupulous engine builders. They modify the correction factor to create a dyno sheet showing the power they promised you. If the multiplier on the dyno sheet is less than one, the test occurred in weather conditions that surpassed the test standard. Conversely, if the multiplier is more than one, then the test day was worse than the standard.

What Is Air?

Any proper discussion of engine-management systems requires thought about the properties of air and the gas exchange process itself. It is commonly thought that air is a single element, and that misconception is further expanded when the word oxygen is used, falsely, as a synonym. By definition, it is accepted that air is 78 percent nitrogen, 21 percent oxygen, and one percent other gases. When we breathe, we take in air, use the oxygen, and exhale carbon dioxide, nitrogen, and the rest. An engine does exactly the same thing. It inputs the total air composition, uses the oxygen, and expels the rest. The oxygen content in air will vary as much as one percent in a normal day and is the result of changes in air density. These changes are caused by variations in temperature and humidity.

An engine needs to mix air with fuel to produce power, hence the term air/fuel ratio. Using the stoichiometric figure of 14.7:1 as an example, we can draw conclusions with the introduction of some basic facts. One gallon of gasoline weighs approximately six pounds, whereas it would take 100 gallons of air to equal one pound. This makes fuel 600 times heavier than air. Therefore, to maintain a stoichiometric ratio, we would need to supply 8,820 gallons of air to burn one gallon of fuel.

Gasoline is a blend of light and heavy hydrocarbons obtained from crude oil. Chemically, it consists of hydrogen, carbon CH_8-18, sulfur, and, in some race gas, lead. The energy content of the fuel is measured by a caloric scale that is read in British thermal units (BTU). One pound of gasoline can deliver approximately 19,166 BTU of heat energy. Like air, gasoline density is affected by temperature. Other liquid fuels such as alcohol have only half the heat energy of gasoline and would require twice the amount to produce the same power. Consider that even the highest-output engine to date is only marginally better in terms of thermal efficiency than the first engine ever produced, using 20–25 percent of the energy that is converted to heat. With this in mind, 75–80 percent of all fuel burned is turned into heat, but not used to produce power; it's transferred to the cooling system and out through the exhaust.

During combustion, the hydrocarbons are broken down and take different forms. The proper name for this cycle is the gas exchange process. The laws of thermodynamics dictate that energy cannot be consumed or destroyed; only its state can change. When the spark plug initiates this event, the hydrocarbons break down into carbon and hydrogen. The hydrogen and carbon mix with the oxygen and the result is heat, carbon dioxide, and water.

3

The ECU and Emissions Compliance

A standard spark plug as seen in the bottom photo offers less of an ionization window than the Accel U-Groove design.

Considering that we all breathe air, the enthusiast must realize a responsible approach to exhaust emissions, accepting that power and exhaust cleanliness can coexist on a street car. Almost every state has implemented some style of emissions testing for licensed vehicles, and a lack of compliance can make your vehicle illegal to drive.

The byproduct of the combustion event in an engine is generally referred to as exhaust emissions. Creating a more powerful engine requires the input of more fuel while concurrently producing more residual end gases, so the enthusiast is caught in a Catch-22 situation. Since the introduction of the Clean Air Act (1970), which was limited to a steady-state idle-emissions test, the requirements have progressed to the current inspection/maintenance (I/M) 240 drive-cycle procedure. With its stricter standards, the drive-cycle program examines an additional pollutant, oxides of nitrogen, which has been identified as a major contributor to ground-level ozone photochemical smog.

The Offenders

The pollutants that we are concerned with are hydrocarbons, carbon monoxide, carbon dioxide, oxygen, and oxides of nitrogen. Each is the result of an inefficient combustion process that can be identified.

Hydrocarbons (HC)

Gasoline consists of hydrogen and carbon atoms, thus any level of hydrocarbon in the exhaust represents unburned fuel. Measured in parts per million (PPM), it is the emissions that make your eyes tear from the exhaust of a poor-running car or an engine with a large camshaft. Mistakenly thought of as a rich air/fuel ratio, HC is known as the misfire meter. HC levels escalate when the engine is not experiencing complete and efficient combustion. HC emissions are created by ignition problems, excessively lean mixtures, camshaft design, excessive

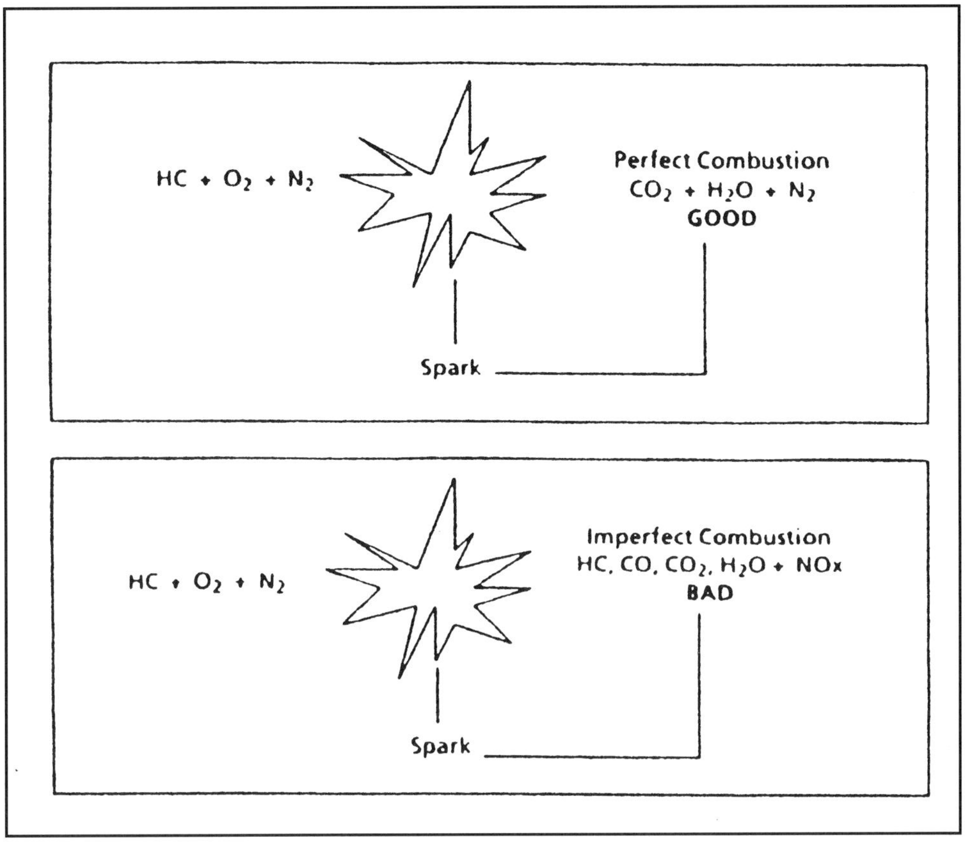

The same elements always enter the combustion chamber. What happens inside determines the amount of pollution produced.

crevice volumes, or carbon deposits in the combustion chamber.

Carbon Monoxide (CO)

Measured in percent, CO is partially burned fuel created by a lack of necessary oxygen to support combustion. Simply put, whenever CO readings are high, the mixture is overly rich. A failed CO test indicates a plugged air filter, excessive injector pulse width, high fuel pressure, dripping injectors, or fuel dilution in the crankcase.

Carbon Dioxide (CO_2)

As a byproduct of the bonding of the carbon and oxygen molecules during combustion, CO_2 demonstrates how efficiently the engine is running. Measured in percent, the higher the reading, the greater the combustion efficiency.

Oxygen (O_2)

Defined as a colorless, odorless gas, oxygen is the most common element known to man. During an emissions test, high levels in the exhaust represent additional oxygen molecules that do not have a carbon element to bond with. Oxygen readings increase if the mixture is excessively lean, and also during scenarios that would normally promote high HC. If the fuel is not being burned, neither is the oxygen. Usually inversely proportioned to CO_2, it is measured in percent.

Oxides of Nitrogen (NOx)

Measured in PPM, NOx is present during all phases of combustion but escalates dramatically when leading-edge flame-front temperatures exceed 2,500 degrees F. Requiring heat, cylinder pressure, and exposure time to be produced, detonation is a leading source of this pollutant. Since 1973, exhaust gas recirculation (EGR) valves have been used to cool down combustion chamber temperatures by introducing inert exhaust gases. At first it may be hard to understand how hot exhaust gas can cool a combustion chamber, but it accomplishes this by consuming space in the bore, thus limiting the amount of combustible mixture that fills the cylinder. Since NOx production is highest under part-throttle light load, when the air/fuel ratio is usually the leanest, NOx was the impetus for the I/M 240 test schedule.

The Catalytic Converter

Anytime a control device is placed in the exhaust stream, it is an attempt to correct what could not be controlled in the combustion chamber.

By definition, a catalyst speeds up a chemical reaction without itself being consumed. Simply, a catalytic converter "scrubs" the exhaust. Early catalysts contained pellets made of 70 percent platinum and 30 percent palladium. The pellet-style converters were eventually replaced by a ceramic monolith of noble metals that exposed a larger surface area while maintainng smaller exterior dimensions. However, not all catalysts are effective on all emissions. Application-specific catalysts are designed to neutralize certain elements; oxidation catalysts convert CO and HC to CO_2 and water; and reduction catalysts treat NOx by converting it back to nitrogen and oxygen. Dual-bed designs are a melding of two different catalyst strategies and affect NOx in

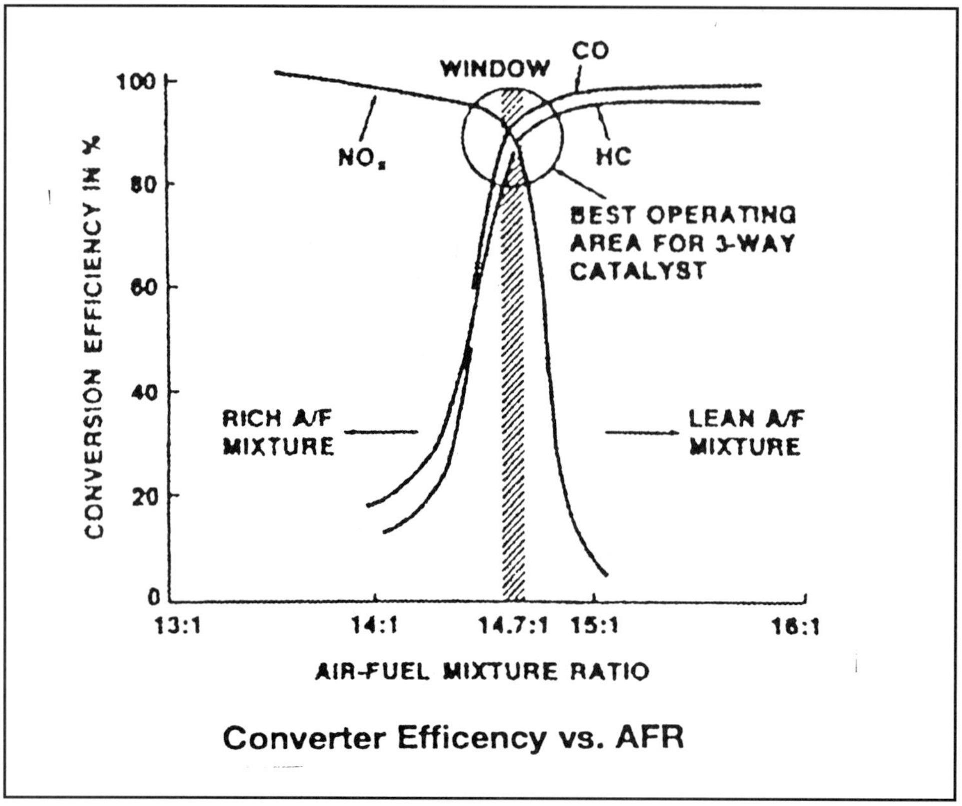

A catalytic converter works only within 2 percent of a 14.7:1 air/fuel ratio.

the first section, CO and HC in the rear section. Three-way catalytic converters simultaneously scrub all three gases in one unit.

For a catalyst to function properly, it must attain a minimum operating temperature of 600 degrees F, but during many driving scenarios they produce temperatures of 1400–1500 degrees F or more. When the conversion process begins, the converter is considered to be in light-off. Converter efficiency is rated in percentage of conversion, 85 percent being the minimum performance standard for today's vehicles. Due to the effects of thermal inertia, which requires a body to become saturated with heat before being able to transfer it, catalysts are placed as close to the exhaust port as possible. Though a definite flow restriction, a correctly sized and properly placed catalyst will impose only a small penalty on power but will have a dramatic effect on exhaust emissions when used with unleaded fuel. Contrary to popular belief, leaded fuels do not plug the converter but rather coat the catalytic material and insulate it from chemical conversion.

Designing an Emissions-Friendly, High-Output Engine

When dealing with combustion efficiency and the gas exchange process, the most challenging emissions to eliminate are HC and NOx. These properties are inextricably linked to the means normally associated with increasing engine output: camshafts with excessive overlap and high compression ratios. Carbon monoxide can be easily controlled with air/fuel ratio adjustments, and O_2 is not a concern since it is used only as a diagnostic gas. Carbon dioxide is tested as an easy means to quantify the integrity of the exhaust system.

When trying to control HC production, the crevice volume, or region, becomes highly relevant. The

In many states, an older vehicle can be modified without any concern for emissions testing.

crevice is an area in the combustion chamber that the flame front cannot access, but one that the air/fuel mixture can. The largest contributor to crevice volume is the area created by the piston and piston rings in relation to the cylinder wall. Other crevice areas are represented by the threads around the spark plug, the space around the spark plug center electrode, and the space between the heads of the intake valve and exhaust valves and the cylinder head, as well as the head gasket cutout.

As the cylinder pressure rises during compression, the mixture is forced into these regions. During combustion, cylinder pressure rises even more, forcing more HC into the crevices. Since the flame travel cannot access these areas, the air/fuel mixture waits in the crevice until cylinder pressure is lower than the crevice gas pressure and flows out the exhaust. For this reason, the trend in current production engines is to move the top ring as close to the crown of the piston as possible, thus curtailing the major contributor to crevice volume.

In an engine that is experiencing proper combustion, it has been determined that the ring package is responsible for 80 percent of the hydrocarbon production, the head gasket crevice 13 percent, and the spark plug threads 5 percent. All other sources contributed the rest. Beside being a major producer of HC emissions, crevice regions waste fuel, which is being used as a filler and not to add energy to the expansion cycle. How intricate does it get? Ford Motor Company has determined that the direction arrow stamped on the piston top was contributing to crevice area, so it was replaced with a mark on the underside of the piston.

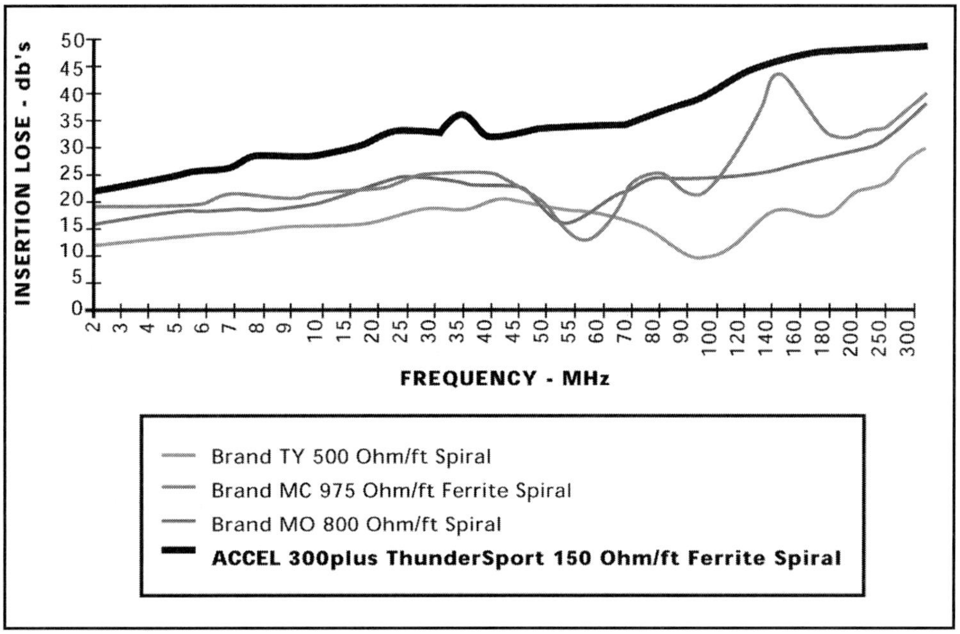

ACCEL ignition wire is designed to get the most energy to the spark plug for the highest power and lowest HC emissions.

Excessive Camshaft Overlap

A camshaft with excessive overlap reduces low-rpm engine efficiency, which in turn promotes poor burn rates and results in high HC output. Overlap should not be considered a completely bad contrivance, though. It can be used as an internal EGR function to pull spent gases from the exhaust port and introduce them into the cylinder. Finding a delicate balance of overlap is ideal but is beyond the realm of the enthusiast. It will remain the domain of the cam companies to acknowledge the emissions issues. Since success is linked directly to a systems approach, if crevice volumes are summarily reduced, then the engine would be more forgiving of overlap because HC emissions are not being stored.

Balancing NOx

Oxides of nitrogen emissions output is the most difficult to balance. Although there is a need for compression ratio for thermal efficiency and to increase burn rates, high cylinder pressure is what produces NOx. The easiest method for controlling NOx is to use a quick-burn-rate combustion chamber. Temperature, pressure, and exposure are necessary for the promotion of NOx, but if the design can speed up burn rates, exposure time would be limited, removing one element from the recipe.

In the 1970s, the domestic auto industry failed to realize that old, slow-burning combustion chambers would not work well enough to satisfy NOx standards, and it drastically reduced compression ratios instead. Rather than redesigning the combustion chamber, it took compression out and added nearly 30 percent EGR dilution. Although EGR is an effective, inexpensive means for controlling NOx, it disturbs burn cycles in the cylinder. The less EGR dilution, the better the drivability will be. EGR is administered during part-throttle, light-load conditions, so it will not affect idle quality or WOT

Many people don't realize that an ignition coil does wear out. Before doing any calibration work, always replace the ignition parts with new ones for the best results.

performance. However, all of this is moot if the engine enters abnormal combustion known as detonation. NOx production goes off the scale during detonation due to the extreme temperature and pressure rise that an uncontrolled release of the end gas energy produces.

Many aftermarket cylinder companies mimic the combustion chamber technology of the latest Detroit offerings. Quick burn rates, internal charge acceleration by means of squish pads, and external mixture motion from swirl of the incoming charge will produce power and pay dividends toward compliance.

Putting It to the Test

The I/M 240 test got its name from the 240-second sampling time with the vehicle situated on a chassis dynamometer to simulate load. The engine runs at idle, is accelerated, and then is allowed to idle again. The tester then averages the results. To certify a vehicle for sale in the United States, the EPA mandates Federal Test Procedure-75 (FTP-75). Divided into three segments, the total test time is

Cool School

If you or someone you know ever wanted to find a place to learn about high-performance engines, your prayers have been answered by Loren Jarvis and the staff at the University of Northwestern Ohio, in Lima (419-227-3141). Working in concert with a standard automotive program, a course of study was added that includes suspension, steering, driveline, engine machining, fuel systems/electronics/ignition, and custom engine building.

Situated on a beautiful 35-acre campus, the technological division encompasses more than 76,000 square feet of classroom and shop space. In addition to a fully equipped machine shop, there is a complete cylinder head flow-testing room with multiple SuperFlow 600 benches, along with Audie Technologies swirl detection and camshaft inspection instrumentation. Airflow theories and machine shop skills are then put to the test in one of the school's two SuperFlow 901 dyno cells, each outfitted with a complete data-acquisition system.

Not limited to engine building, the high-performance drivetrain and chassis programs teach the essentials of these theories. Subjects include setting up rear end gears; adjusting and building a chassis for drag, dirt, and road racing; and tuning a four-link for better 60-foot times.

The university acknowledges that to be successful, skills beyond automotive performance are required. Addressing these needs, the associate's degree program covers mathematics, English, psychology, contract law, and introduction to business. The staff comprises hands-on instructors who collectively have more than 826 years of practical and theoretical experience. More important, they are hot-rodders at heart. This creates a common bond with the students that is not present in most institutions of higher learning.

Part of the curriculum is to gather practical experience around racing, so in an unprecedented move, Northwestern purchased nearby LimaLand Motorsports Park in 1997. At this location, student activities are at their best. Associate Dean Bill Sergent states, "Owning our own facility, where the students can work with our educators in an environment that includes a drag strip burnout pad for 60-foot shots and a complete quarter-mile dirt track and road race circuit, allows them to gain invaluable practical experience."

The school has an open admissions policy and accepts students who have earned a high school diploma or equivalent. There are no limits on age or past experience. Their mission statement is to offer quality diploma, associate, and baccalaureate programs to traditional and nontraditional students. Because the university is an accredited institute of higher education, tuition assistance is available with loans, grants, and work-study programs. Prospective employers acknowledge the prestige of Northwestern graduates, with students gaining employment at Mr. Gasket Company, Inc., Summit Racing Equipment, and Skip Barber Racing School, along with major NHRA, NASCAR, and IRL racing teams.

A Generation VII ACCEL ECU is running this Ford industrial engine on propane. This system produces virtually no pollution and can be used indoors.

4500 seconds (75 minutes) and is performed on an extremely sophisticated 48-inch roller chassis dynamometer with a load cycle designed to simulate morning commuter traffic in Los Angeles. Since 80–85 percent of the engine emissions occur during the first two minutes of operation, an independent cold-start test is also performed with the vehicle placed in a room and chilled to 20 degrees F.

In another procedure called SHED, the vehicle is placed in a test cell that resembles a paint booth. The HC emitted while the engine is not running is measured for 24 hours. While state I/M programs measure readings in either percent or PPM, the FTP cycle weighs the tailpipe output and records grams of each pollutant per mile of operation. For this reason, smaller-displacement engines are easier to certify since they displace less air. And even if they are dirtier, they'll exhibit fewer recorded grams/mile. Some states will initially adopt a shorter version of I/M 240 called 5015. The difference is that it calculates 50 percent of the load while accelerating to 15 miles per hour.

Clean Combustion Chambers

Every combustion process, even the most efficient one, creates carbon that becomes deposited on the combustion chambers of the cylinder head and on the crown of the piston. An additional area of concern about deposit formation is the back side of the intake valve and the intake track of the manifold during valve overlap and EGR operation, if the car is so equipped. These carbon-based deposits are detrimental to the engine, as they raise the octane requirements when found on the piston and combustion chamber, and absorb fuel when attached to the intake valve.

Many old-time mechanics used to carefully drip water down the carburetor with the engine running at high idle in the hopes of loosening and removing the combustion chamber formations. Others believe that using a drip of automatic transmission fluid applied with the same procedure as the water does the trick. Both methods have some intrinsic value for carbon removal, but their effectiveness is limited and, at times, costly. An uncontrolled drip of any liquid into an intake system of an engine affords the opportunity to hydrostatically lock the engine if too much is administered. Since a liquid is noncompressable, the connecting rod or piston gives instead. A concern with using water or transmission fluid to loosen deposits is rooted in the fact that the carbon is not loosened, but hammered, from the parts.

Recognizing that there had to be a better way to remove deposits without dismantling the engine, C.A.T. Products (215-639-1839) developed the Run-Rite process to chemically dissolve both combustion chamber and intake valve deposits and let them burn off through normal combustion. A complete kit is offered for less than $50, which includes a reusable applicator. A three-step process has one part added to the gas tank, another to the engine oil, and the third and final element safely dripped into a manifold vacuum source. The Run-Rite applicator is designed to limit the liquid throughput of the company's patented carbon-dissolving chemical. Simply connect the applicator bottle to the appropriate vacuum source, set up the idle speed, and in about 30 minutes the deposits will be removed, lowering the emissions and octane requirements and restoring full performance.

The ECU and Its Role in Emissions Compliance

To date, all ACCEL ECUs, like a good majority of conventional speed parts, are considered illegal for street use when following the EPA to the letter of the law. When viewed in the light of passing a tailpipe emissions test, the ACCEL system is more than compliant when tuned properly. Since the 6.0 ECU does not have the ability to control EGR, air injection reaction (AIR), or charcoal canister purge functions, it is not considered acceptable by the EPA. Even with the ECU's ability of total fuel and spark control, it would be very difficult to tune a modern vehicle to pass an emissions test if a camshaft with excessive overlap is installed. Increases in overlap will create high hydrocarbon emissions and will also have an effect on the oxygen sensor output at low engine speeds due to

1-PULSE AT CRANKSHAFT REVOLUTION

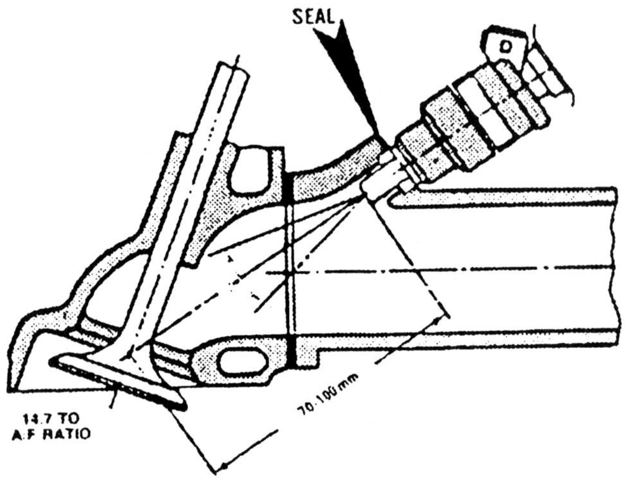

Typical Fuel Injector Installation

Simultaneous double-fire coding of the ECU means that the injector will fire once for each crankshaft rotation, and twice for each distributor rotation.

reversion and falsely introduced oxygen. In many instances, even though engine-out emissions numbers may be high, the use of an original-equipment catalytic converter and its high conversion rate will do an excellent job of cleaning the exhaust.

ECU Options

Both the standard spark/fuel and the SEFI ECUs have a number of options available to enhance their operation. All options that are available could be installed at the time of the original purchase or added later as the needs of the management system change. The following discussion is in regard to actual hardware options, not changes in configuration through the CalMap software.

The majority of spark/fuel ECUs externally resemble each other whether they are designated for Buick, Ford, or Chevrolet usage, port or throttle-body injection, with either OE-style ignition or inductive pick-up; they are not interchangeable and have major internal differences for ignition compatibility. In the case of throttle-body injection, a major internal difference is found in the firing sequence of the injectors. The simultaneous double-fire (SDF), sequential, and marine ECUs also use different enclosures and are visually different, along with their obvious internal changes.

The 015013 wire list pertains to all spark/fuel ECUs regardless of vehicle make designation, whereas the marine and SEFI sharing a dedicated wiring harness.

The internal changes that are made to a standard spark/fuel ECU are limited to ignition compatibility. The simplest way to think of ECU designation is to identify the type of ignition being used. Since the 015013 is coded for simultaneous double-fire operation, the ignition firing order becomes moot and gives this ECU wide-based installation appeal. It is very common to see this ECU running fuel-injection systems on imports, domestics other than Ford or GM, and with industrial engines.

To begin the identification of the ignition requirements with a GM or Ford application, you need to decide if you are going to use a standard OE-style electronic timing control distributor. This design has no provision for either mechanical or vacuum-advance mechanisms. If electronic control of the timing is going to be used through the above-mentioned OE systems, then ACCEL part numbers pertaining to standard Chevy or Ford ECUs will apply. The differences with the Chevy ECU listings (part numbers 74021, 74022, 74023, 74024) all refer to fuel and spark calibration changes for varied cubic-inch engines.

If ECU-controlled timing is desired but there are no electronic spark timing distributors that easily interface with the ACCEL ECU, such as Mopar, Honda, and industrial engines, control of the spark timing can still be accomplished with the use of an inductive pickup, either in the distributor or on a crank sensor. In addition, the interface of an after-market ignition amplifier such as the ACCEL 300+, MSD, Mallory, etc., will need to be included. For this use, ACCEL ECU part number 74042 would need to be ordered, which is a standard ECU with a board modification to accept the inductive pickup signal. If you currently own a standard spark/fuel ECU and you want to have it converted from an OE-style timing control to inductive pickup, it needs to be sent back to ACCEL for board modification 74043-1. Once an inductive pickup modification is installed, the ECU becomes nondescript and is not designated as a Ford or Chevrolet unit.

ACCEL part number 74025 is a Buick GN spark/fuel ECU and is coded differently for the recognition of the Hall effect crankshaft and camshaft sensor signals. An inductive pickup generates a modified sine wave identified as a sawtooth, while the signal from the Hall effect is a square wave. The Buick ECU modification consists of an internal PROM and software change.

Chevy LT1 ECU—The Chevrolet LT1 ECU part number 74022-L also consists of a board change to interface with the output of the Delco Opti-Spark ignition. If the primary triggering function is to be moved out of the Opti-Spark and an IPU crank trigger is employed, then a 74042 needs to be ordered.

Marine ECU—Also offered is a marine ECU that consists of a 74022 in a weather-tight enclosure with weather-tight bulkhead connectors. This ECU is not only used in marine applications but is commonly found in late-model Corvettes that require the ECU to be mounted under the hood. Other uses for this product are industrial or severe-environment duty.

SFI ECU—Sequential spark/fuel ECUs that are offered to interface with OE electronic timing controls are applicable only to Ford TFI and Buick GN configurations. All other SEFI ECUs are designed to work with IPU. This style of ignition is also available for Ford ECUs. Other than ignition compatibility, all SEFI ECUs have dedicated injector harnesses for proper phasing of the injectors.

TBI ECUs—ECUs for ACCEL throttle-body units are coded to fire the injectors in a different way from a conventional ECU. If electronic timing control is desired for a throttle-body unit and an electronic spark-control distributor is not offered, then an IPU modification must be performed.

Stricter emissions laws are forcing industrial engines to convert to EFI. This airport tug is now DFI- and Ford-powered.

ACCEL makes ignition wiring harness adapters for most new domestic engines to plug into their high-powered ignition systems. Shown is a Dodge Magnum truck harness.

Interface Harness—Pioneered by EMIC Dan White, and now also produced through other sources, jumper interface harnesses allow the connection of a 015013 ECU to either a GM or Ford OE harness. The benefit of this style of installation is the ease of connection with no cutting of the car's original harness, and the ability to change back quickly to the factory engine-management electronics. The down side of this installation is that in many cases the jumper harness is wired so as to not use all of the drivers in the ACCEL ECU. This limits the range of injectors that can be operated.

A very popular option is part number 7043-I, which allows the use of an inductive pickup distributor or crank trigger as the primary ignition switching signal. The benefit of a crank trigger is the accuracy in identifying the actual piston position in the bore from the crankshaft. This is in contrast to a distributor with an inherent stack-up of tolerance from the timing chain and its intermediate components. This option also allows the use of ECU-controlled programmable ignition timing on engines that did not originally incorporate an electronic spark timing distributor, or were not ignition-compatible with the ACCEL ECU.

The Ford TFI is considered an E-coil, for epoxy coil. This new technology is in contrast to older, oil-cooled ignition coils.

User Interface Module—The acronym UIM is used to describe a stand-alone interface that allows the end user to alter the air/fuel ratio that is programmed into the core fuel table both richer or leaner at idle and wide-open throttle. The module has two knobs: one for idle, one for WOT. Simple to install, the four leads plug into the ECU bulkhead connector P1 at terminals C6, D4, C4, and D8. The UIM has no control above one percent throttle opening and below 99 percent TPS. This is an excellent option for the end user who likes to do some minor tuning and is not interested in mastering the CalMap software.

Linear Oxygen Sensor Kit—To install this kit, the ECU needs to be sent back to ACCEL for a PROM change. The linear oxygen sensor kit includes an NTK-sourced wide-band module (known as "the Brick") and an NTK 5-wire oxygen sensor that replaces the standard-issue Bosch sensor. All necessary wiring is included to connect the components to the main wiring harness. The linear feedback option allows for multiple target air/fuel ratios for total feedback control and operation.

This is in contrast to the standard ECU, which, if operated in closed loop, has a target mixture of 14.7:1 as its only set point and is programmed to correct to that target only under certain operating conditions. The linear option will allow target air/fuel ratios referenced from every two rows of the fuel matrix based upon engine load, providing eight target air/fuel ratios. This option also incorporates the ability to set a target warm-up air/fuel ratio that will be maintained after the initial 45 seconds of running and until the closed-loop temperature parameter is met in CalMap. Deceleration "leaning" through coast-down fuel shut-off monitors TPS and RPM and offers programmable set points to shut down the injector pulse width during closed throttle for better emissions and fuel control.

Altitude Adjustment—ACCEL offers an onboard change to the ECU and an additional one-bar MAP sensor that is referenced to atmosphere to function as a correction to the base fuel map for changes in barometric pressure. This option uses the same ports in the bulkhead connector as the UIM and can not be used in conjunction with this option.

3-Stage Nitrous—The standard spark/fuel ECU has the ability to control one stage of nitrous with the optional ability of adding two more stages. For this to work, ports P2-B5 and B4, which originally controlled the coolant fan and TCC functions, must be sacrificed.

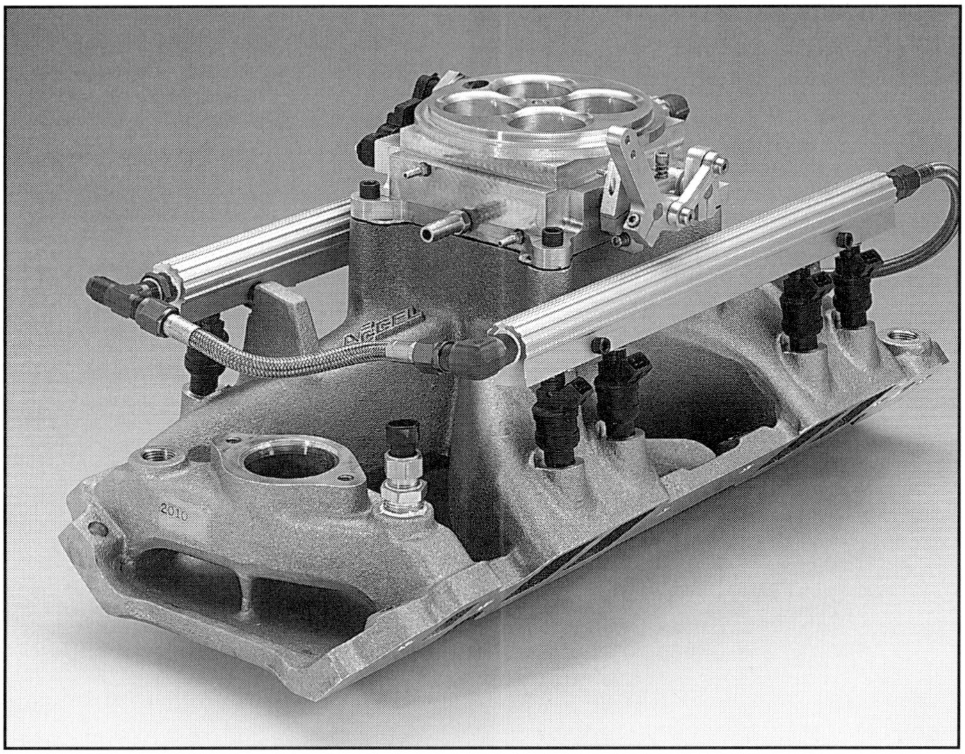

4

Matching the Intake Manifold to the Engine

The intake track of an engine and the camshaft profile are most responsible for the engine's personality, described as the rpm range for both peak power and torque. This does not occur by coincidence but through physics.

A factory EFI engine is usually designed around torque production, with the reasoning that even though we buy horsepower, we drive torque. Torque can be loosely defined as the amount of work an engine can do, whereas horsepower measures how quickly the work is accomplished. Since horsepower is a mathematical derivative of torque, engine speed or RPM is required for big horsepower numbers. An excellent example of this is a large diesel engine such as those produced by Detroit Diesel, Cummins, or Caterpillar. Since diesel fuel burns very slowly, the RPM potential of a compression ignition engine is limited, with most over-the-road trucks having anywhere between 300 and 500 horsepower. An automotive enthusiast may scoff at these numbers, saying that a good small-block can make more than that on pump gas. But what that statement ignores is that the diesel's power is produced between 1500 and 1800 rpm. Mathematically converting 500 horsepower at 1500 rpm to torque, the engine is producing a whopping 1751 lbs-ft., and that is what gets the load down the road! Detroit has found favor in torque-biased engines for many reasons, but mainly because they can use a smaller engine to move a heavier vehicle while enjoying the fuel economy and lower emissions output of fewer cubic inches. Most do not acknowledge that as an engine grows in displacement, so does its emissions. The EPA weighs the amount of emissions emitted from the tailpipe in a scale of grams per mile, and the larger the engine, the more grams of pollutants are discharged. In addition, a torque engine will allow the use of a low numerical drive gear ratio and a tight torque converter, all attributes for fuel efficiency. Likewise, and important to an OE calibration engineer, tip-in throttle response is better with a torque-biased design.

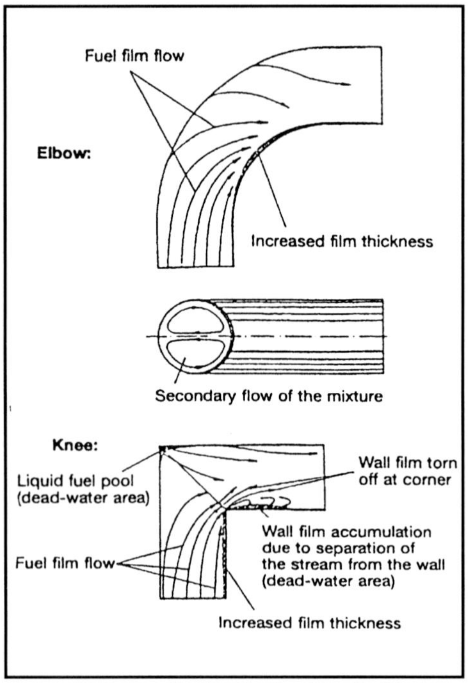

Neither air nor fuel mixed with air likes to make turns. The most efficient intake manifold designs have as straight a path to the cylinder head as possible.

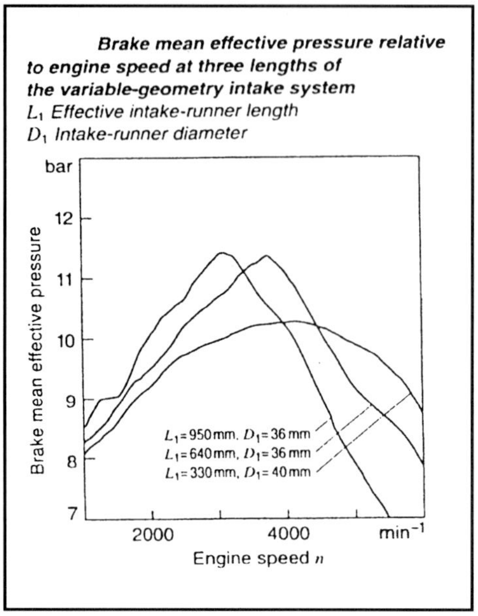

Engineers gauge brake mean effective pressure (BMEP), not horsepower or torque, as a true measure of an engine's potential. The intake manifold runner length determines the engine rpm at which peak cylinder pressure will occur.

ACCEL dealer Rance Baxter converts many unusual applications to EFI. Here is a Chrysler 426 Hemi with a modified BBC Super Ram manifold.

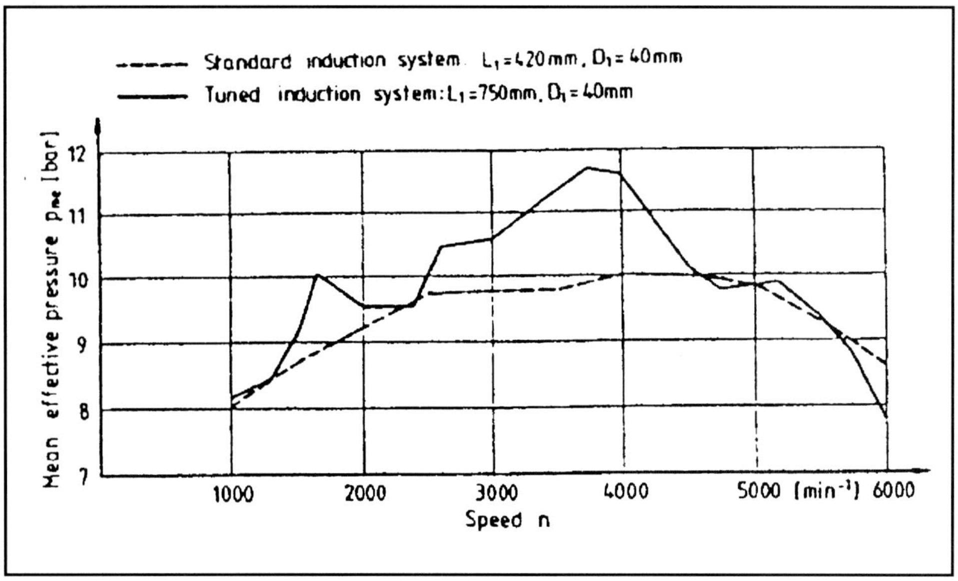

A tuned intake system will create a defined tuning point. An untuned intake manifold will produce less peak torque but will have more area under the curve.

Intake Manifold Tuning Theories

An internal combustion engine can be described as an air pump, and during this process a pressure wave phenomenon is created in both the exhaust and intake tracks. When referring to intake manifolds, this action is identified as resonant tuning.

The most common mistake that engine builders and enthusiasts make in modifying EFI engines, especially the GM TPI, is choosing the wrong camshaft. The problem stems from the fact that most people are not familiar with inlet manifold tuning and the relationship it has with camshaft selection.

This is nothing new and has been the case since the first hot rod tore down an abandoned air field in post-World War II America. With fuel injection and long intake manifold runners, as on TPI and Ford 5.0 in particular, engines become much more sensitive to camshaft selection.

Airflow Characteristics

As with any technical discipline, more than the direct issue should be understood. It is impossible to discuss intake tuning and its relationship with cam profiles without being conversant in airflow characteristics and the tools used. This chapter is meant only to be an overview and is by no means all-inclusive. People have spent their lives studying these areas. My goal is to help you begin to understand how to select an intake manifold for EFI and the necessary camshaft to complement it.

An engine is a pump. The more air it moves with the proper amount of fuel, the more power it makes. We are

ACCEL understands the relationship of the camshaft profile to the intake manifold runner length. The company offers grinds for TPI GM engines in both flat-tappet and roller-hydraulic styles.

The ACCEL Street Ram is a popular conversion for GM pickup trucks, with its huge bump in torque from the factory TBI.

all aware of the little things we can do to increase airflow in an engine. Open-element air filters, larger throttle bodies and the like usually increase performance. Why? What has been actually done? These items have increased the airflow potential to the engine, hence its power output is greater.

In practice, the study of airflow consists of the path the air must take from atmosphere (the point at which the air filter draws its charge) to its final point, where the spent gases reenter the atmosphere at the tailpipe. For the purpose of this book, we will mainly be concerned with the incoming air charge path to the cylinder.

GM's TPI System

Most of us remember the first time we saw a GM TPI engine. We were most likely struck by the myriad tubes and manifolding and its racy appearance. If looks were revs, a TPI would spin to 7000 rpm easily. But it did not work that way. It developed power, but in a different manner. Chevrolet engineers reconfigured the delivery of power and torque. It sure felt good when we hit the gas pedal, but we could not help but wonder why all the power went away by 4500 rpm.

Some of us associated all those racy-looking components with the requirements of the injection system. Nothing in the entire intake system resembled parts of a carbureted engine. But why did this look so different when an engine does not care how fuel gets in its cylinders? Most OE EFI manifolds look different from their carburetor counterparts because of the opportunities that injecting the fuel at the intake valve gave engineers.

An EFI system uses a dry-flow method of air/fuel delivery. Dry flow means that only air courses through the plenum and runners. Fuel is introduced at the base plate under the runners, aimed directly at the back of the intake valve. Carburetors mix fuel and air together at the beginning of the induction stream, and the intake manifold brings an already measured air/fuel ratio into the cylinder heads.

By using a dry-flow manifold, engineers have the opportunity to drastically increase intake runner length and, in turn, increase and change the torque output of the engine. No longer was the distance between the two cylinder heads an obstacle in tuning the induction system to deliver the type of torque that engineers wanted. By not having

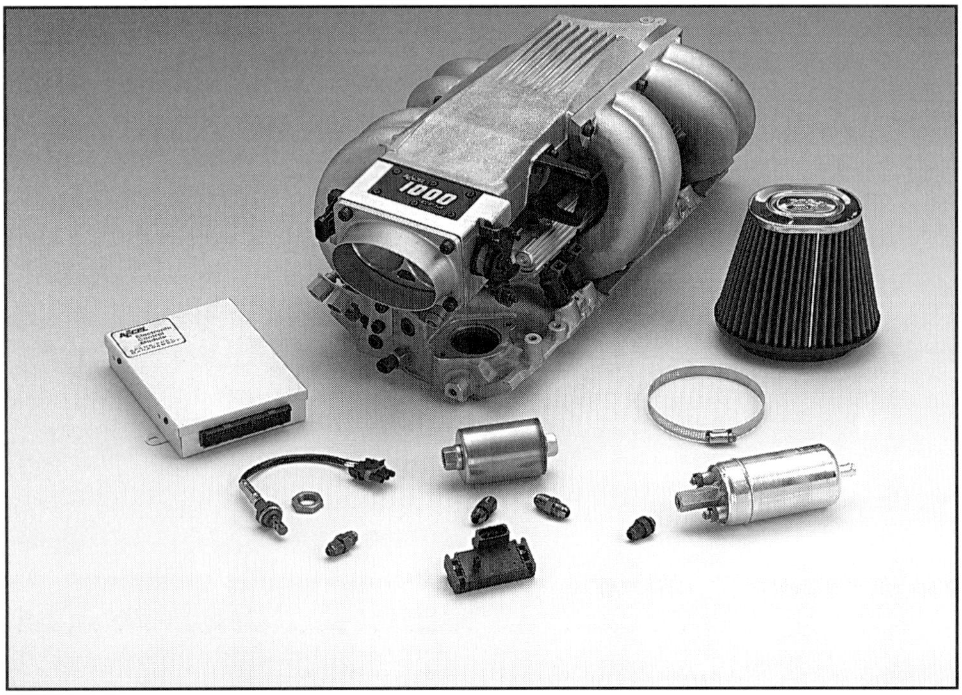

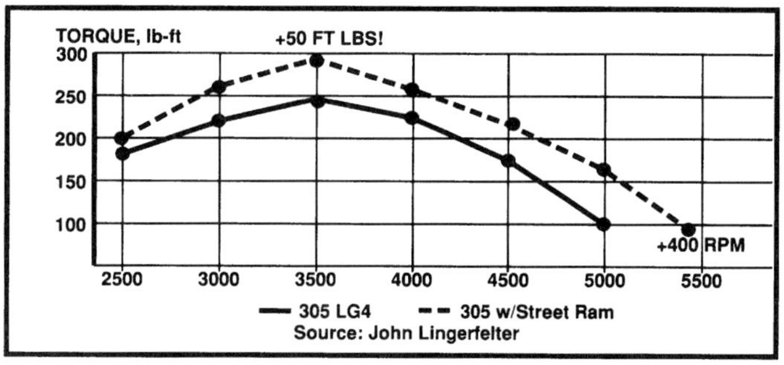

The numbers tell the story: The Street Ram is more than 50 lbs.-ft. of torque stronger on a 305 than either a carburetor or a GM TBI.

to transport fuel with the air, runners can be made to twist and turn to acquire the desired length.

If engineers tried to do this with a wet-flow manifold, they would experience fuel dropout and puddling from all the turns. The closest thing to a TPI motor's runner length was accomplished by Chrysler in the late 1950s and early 1960s. They had a long-runner cross-ram intake manifold that placed two carburetors (one on each side) outboard of the valve covers on its big-block V-8. That positioned one carburetor by the headlight and the other by the master cylinder, on opposite sides of the engine, to achieve the desired resonant tuning. This design was short-lived since the long runners and wet-flow style offered poor driveability at small throttle openings and low velocity.

Intake Manifold Design

Common parts to original-equipment intake manifolds include a butterfly (to control airflow), a plenum (common area to all cylinder runners), and a runner to feed each cylinder. The ideal intake manifold would offer low airflow resistance, good cylinder-to-cylinder distribution, sufficient but not excessive heating (on wet-flow designs only, for vaporization of the fuel), and runner lengths to take advantage of tuning.

One of the equations used when designing a runner length for a given engine is called the Helmholtz resonator. This equation can predict the approximate rpm at which increased airflow will occur due to intake tuning.

When an engine is operating at the rpm at which the intake track is most efficient, a ram tuning effect takes place, called inertia supercharging. When this occurs, an extra air column is actually shoved into the cylinder due to the length of the intake runner, effectively raising the volumetric efficiency, or the amount of cylinder fill. That is why many modern engines have dual-path intake manifold designs. By using internal butterflies, the rpm that the manifold will resonate at is controlled to expand the power band. Long runners are used to enhance low-speed torque, and short runners feed the cylinders at high engine speeds. Because the GM TPI intake manifold is a tuned design and there is a fuel injector for each port, it was named Tuned Port Injection.

Air Path

Before the air can enter at the intake valve in the cylinder head, it must travel a great distance and make numerous turns. Even though a dry-flow manifold is much more forgiving than a wet-flow design, air still does not like to make turns. Whenever you ask air to change direction, velocity suffers, and in turn, flow rates drop.

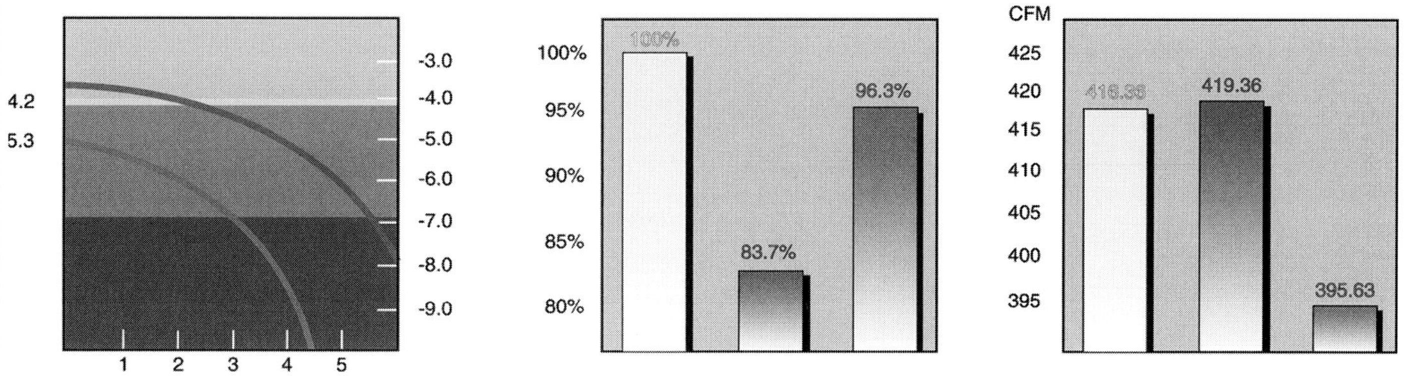

The ACCEL PowerFilter produces more cylinder pressure than the most popular brand of high-performance air filters.

When looking at a TPI, keep in mind that the left runner feeds the right cylinder head, and vice versa.

As a comparison, let us compare the runner length of Edelbrock Performer (dual-plane) and Victor Jr. (single-plane) manifolds vs. a stock TPI.

Minimizing Heat—Common practice among drag racers with EFI systems is to ice down the plenum and runners before each pass. Most racers have found empirically that when a dry-flow manifold is hot, there is a great loss of performance. This is a function of both the absence of the latent heat of vaporization and the fact that the intake manifold runner in many cases is extremely long. Because the air has so far to travel from the throttle body to the intake valve, and because it is not carrying fuel as a coolant, it has plenty of time to heat up. For every 10-degree F increase in charge air temperature, power output drops approximately one percent. The low hoodline of many new cars often exaggerates this problem by having the hood close right on the plenum. There is simply nowhere for the heat to go except into the intake components.

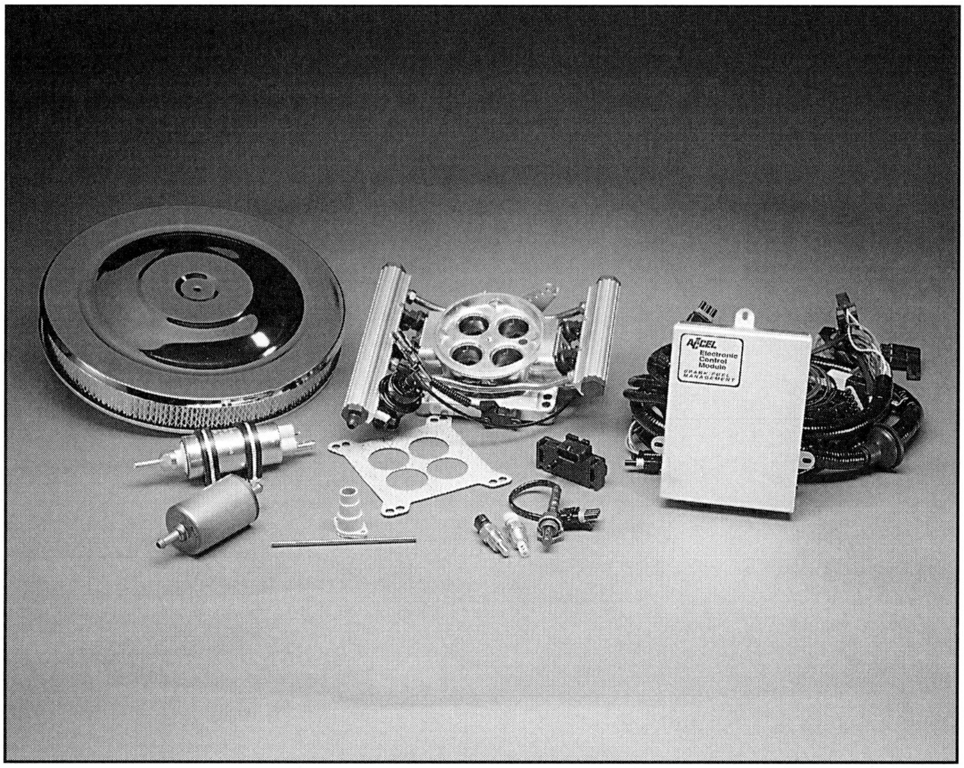

ACCEL TBI systems are of the 4-barrel design and fit a conventional performance intake manifold bolt pattern.

It is critical to try to minimize heat transfer to the intake charge. That is why ACCEL uses a manifold air temperature sensor to correct the air/fuel ratio as the air density changes through temperature fluxes.

A very effective way of insulating

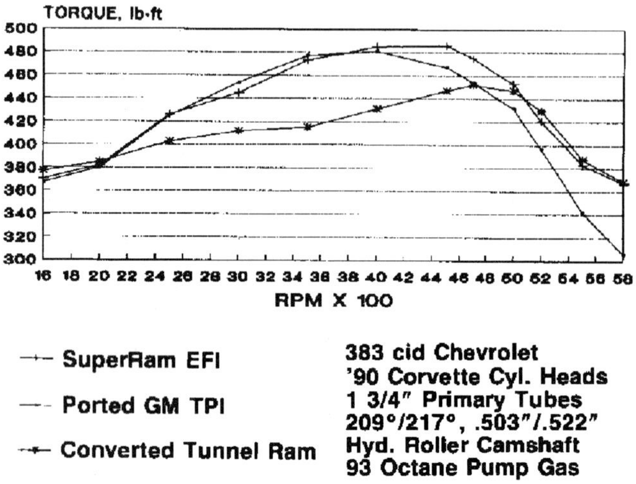

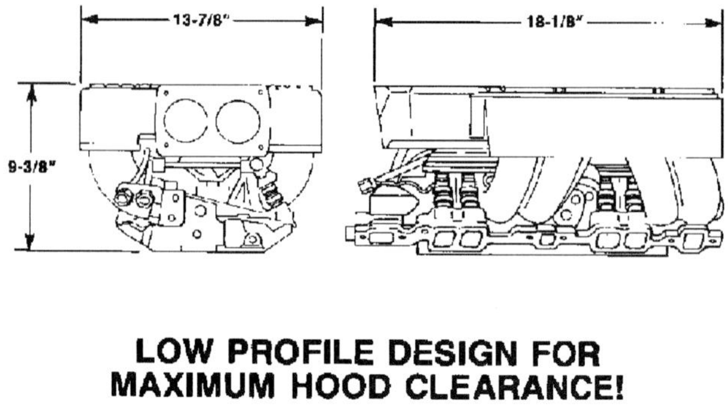

The Super Ram for small-block Chevrolets offers a low clearance to fit C-4 Corvettes. Because of this, the entry angle into the cylinder head is less than ideal.

the intake manifold and runners from heat is through coatings. By applying a thermal barrier coating on the bottom of the intake manifold, the heat transfer from the engine is reduced. In contrast, a thermal dispersant on the top of the intake, runners, and plenum lets heat escape from these components and decreases power loss.

As important as it is to be able to feed air into an engine, it is equally imperative that we allow the cylinder to exhaust itself. When optimizing a package, a minimum of 70 percent intake/exhaust ratio should be accomplished. The present trend is to be at 80 to 85 percent intake/exhaust ratio, especially if it is going to be a high-rpm engine. Most modern production engines usually run best a 80 percent. Any gains in exhaust flow above that are usually harder to obtain than the net gain in power is worth.

Intake Manifold Testing

When either designing or modifying a component that moves air, a flow bench becomes an essential tool for testing and tuning. A flow bench allows the user to accurately measure the amount of air (in a scale of cubic feet per minute) that will pass through the given test piece. Since airflow in an engine is unpredictable, it is extremely important to confirm that what you are doing is actually increasing airflow and not detracting from it. Many times a part appearing to have better airflow characteristics actually tests worse than the component it might be replacing.

Velocity probes are used in conjunction with the flow bench to check the speed of the air in a given area. By reading a dedicated manometer and using a conversion chart, the tester is able to determine the charge's velocity. Remember, air does not like to turn. Just as a car would skid traveling too fast around a corner, so does air. When air skids, it separates or shears. This is a very undesirable phenomenon because, at the point of separation, the effective area of the port is diminished.

Molding Rubber—Molding rubber, commonly known as cold pour, is used to make molds of the insides of ports. This allows you to see the actual path the air must take, and, therefore, look for obvious restrictions. Airflow experts use this method extensively to aid them in improving their designs.

As good as flow bench data is, it shows only potential for airflow. It does not measure inlet manifold tuning characteristics (at which rpm the resonance takes place) or factually simulate the flow in a running engine.

Gathering Flow Data

For demonstration purposes, I have taken several components of both stock and aftermarket GM TPIs and flowed them on a bench. The results can be seen on page 104. The criterion for the valve lift numbers was taken from the lift of a stock 1989 Corvette camshaft at the stock 1.5:1

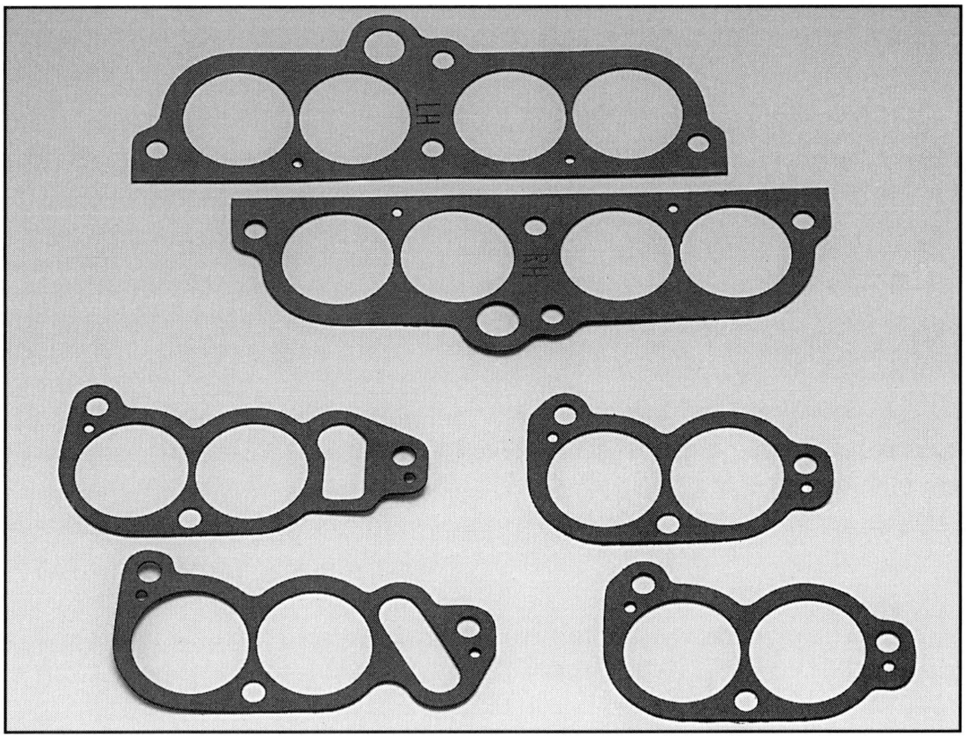

Complete gasket service kits are offered for all ACCEL TPI intake systems.

A service kit is offered to keep your air filter performing like new for years to come.

The big-block Chevrolet Super Ram system has found favor with enthusiasts both in water and on land.

rocker ratio. Since it is common to fit 1.6:1 rockers, I also flow-tested at that resultant lift. The next test used the same components, but at the lift of the popular ACCEL 74211 camshaft, and then again with both 1.5 and 1.6 rockers. I also tried to put together any and all combinations of stock and aftermarket parts that readers may have on their cars. A stock 1989 Corvette aluminum cylinder head was used as a test bed. The flow data for the cylinder head alone is used as a comparison of how efficient each intake system is. The equation:

Cfm @ 28 inches H_2O x .257 x the number of cylinders

was used to show the power a given airflow could support in a completely optimized engine that limited all losses. However, to make that power, everything has to work in unison. The engine needs the proper compression ratio, camshaft, rod-to-stroke ratio, etc. It is what I call a backdoor equation. If you know the horsepower you are looking for, you can calculate the needed airflow. Likewise, these charts show flow data alone and do not take into consideration any of the benefits of a more aggressive camshaft other than its increased lift. Duration, lobe

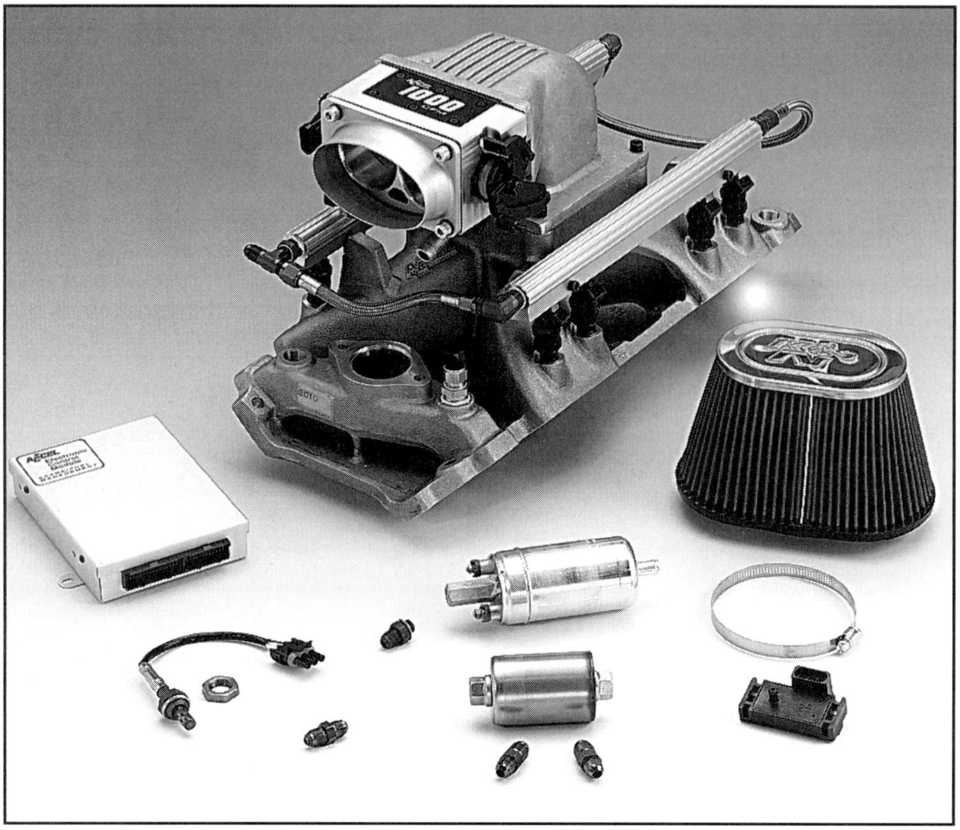

The ACCEL single-plane intake manifold can use a throttle body adapter if the installation so requires.

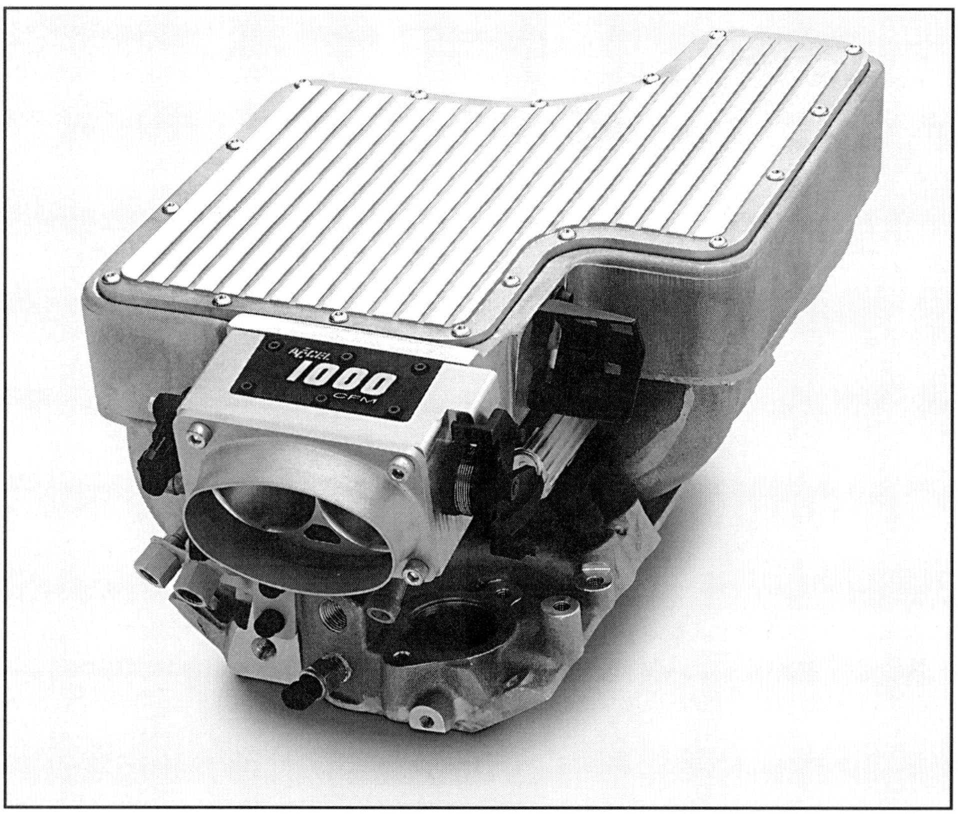

Known to many as the "pizza box," the SBC Super Ram is still the best TPI manifold more than 12 years after its introduction. It works best on engines up to 406 cid.

separation angle, valve opening angle, and ramp are not calculated.

The length of the runner is also not taken into account. For example, the short Super Ram runner will allow the engine to spin faster to make more horsepower than a longer runner, but at the expense of less torque at low rpm.

I also included the flow data for a completely ported cylinder head of the same design. It is interesting to see how much the manifold affects a more efficient port than stock.

Please note that this is not an intake comparison. Rather, it's solely an example of how different components will work together. It can also be used to explain why your engine may not be performing to expectations.

Interpreting Flow Data

A completely stock GM TPI has the ability to flow enough air to support approximately 329 horsepower. By increasing the lift alone with a 1.6:1 rocker, the peak airflow can then support 335 horsepower. Switch over to ACCEL Hi-Flow runners in conjunction with the 1.6:1 rockers, and there is potential for 347 horsepower.

Note that a completely ported cylinder head, when attached to a stock TPI, flows marginally more air than a stock head with ACCEL Hi-Flow components and, in turn, has almost the same power potential. The short runner of the Edelbrock carbureted intake has the airflow potential to support slightly more horsepower than an ACCEL large-tube runner combination. But keep in mind that horsepower is not everything.

Torque is a key factor, along with what is known as area under the curve.

Sidebar: Camshaft Terms

Understanding camshaft terms will you allow to pick the grind that will work best for your application. The following is a brief description of cam lobe and design terminology.

Ramp—The portion of the cam lobe event from zero lift (base circle) to the defined opening or closing point.

Base Circle Radius—The portion of the cam contour at zero lift (valves closed).

Flank—The section of the cam lobe between the ramp and the nose radius.

Nose Radius—The radius that is tangent to the flank and the maximum lift point. Also, the instantaneous radius at the maximum lift point on the nose of the cam.

Inflection Point—The point on the lift curve where acceleration (of the lifter) changes from positive to negative, or vice versa.

Main Event—The portion of the cam lobe profile from the ramp to the nose. There are two main events, one each for the opening and closing halves of the profile.

Velocity—The rate at which position changes. In most instances, it is expressed in terms of distance vs. time (mph, feet per second, etc.). When dealing with camshafts, time is replaced by degrees of camshaft rotation. This allows the study of camshaft profiles to be independent of the engine speed.

Acceleration—The rate at which velocity changes. It is expressed in cam logic as inch/degree or thousandths/degree.

Jerk—How fast the acceleration changes, expressed in units of inch/degree.

Duration—The amount of time in crankshaft rotational degrees that it takes for the lifter to travel from a specified lift point on the opening side of the lobe to the same point on the closing side.

Lobe Centerline—An imaginary line referenced from the true center of the base circle out to the end of each lobe, expressed in crankshaft rotational degrees. Intake centerline values are found in the crankshaft's rotation ATDC, whereas exhaust centerlines are found at travel BTDC. Lobe centerlines are commonly thought to be the actual midpoint of the flank and are usually close to that, but different ramps on the opening and closing sides may change that reading.

Lobe Separation Angle (LSA)—Measured in camshaft degrees, a reference to the amount of rotation it takes to travel from the intake centerline to the exhaust centerline. As this number increases, the distance between the centerlines is spread out. LSA is often considered an indicator of overlap, or the period of time both values are open on the same cylinder. Because it is a partial function of LSA, overlap can be calculated with fair accuracy using duration and LSA values.

The BBC Stealth intake manifold is a long-runner torque design that works great for towing, or in heavy vehicles.

Ideally, you want a powerband as broad as possible, making the vehicle the most responsive. Exposing the ACCEL components to the Extrude Hone abrasive flow media process allows them to move enough air in conjunction with the ported cylinder head to support 429 horsepower at a conservative 0.533 inch lift. See the charts on page 104 to find your combination and see what its potential is.

Cam Profile

Most hot-rodders think that more is better. From my experience, if you have disappointing results after installing a cam in an EFI engine, it can usually be traced back to the duration. By increasing duration, you effectively raise the rpm range. It is critical that the camshaft work in the same range as the intake manifold. If it doesn't, the engine will have a very narrow power band, if it makes any power at all.

With a stock torque converter and any one of the long-runner intakes, it is best to use a duration of no more than 208 to 220 degrees at 0.050 inch tappet lift on a 300- to 350-cubic-inch

This adapter hat is offered under part number 74882 to use a GM TPI-style throttle body on an intake manifold.

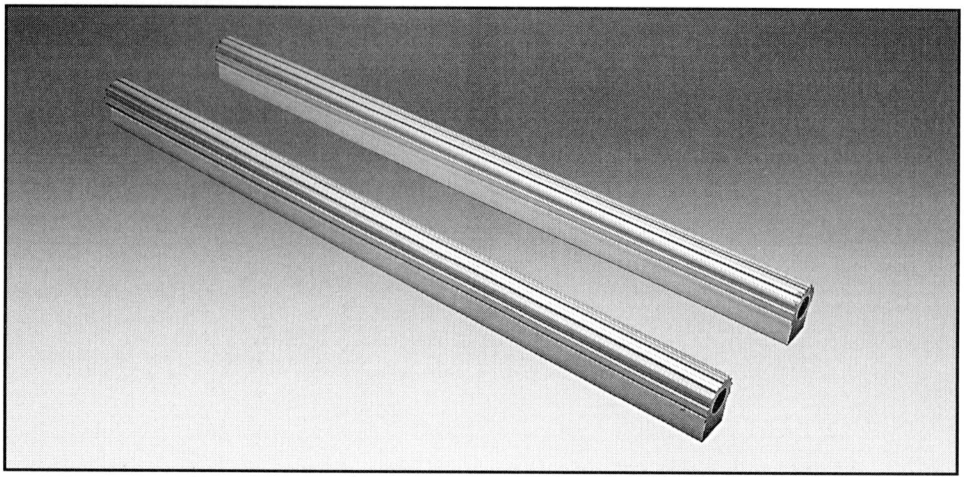

ACCEL can supply everything you need to do your own EFI conversion of a carburetor intake manifold, even raw stock fuel rails.

engine, with around 208 to 210 degrees preferable. Excessive duration will cause resonance back into the plenum and result in surging and rolling at idle, with no gain in power and a loss of bottom-end driveability.

Split Duration Cams

A split duration cam (different durations from intake to exhaust) is most effective with less than 80 percent intake/exhaust ratio. If the cylinder head has a ratio better than this, the duration can be similar or close to the same on both lobes.

Choosing Lift

When settling on the amount of lift to use, choose a value that is near the stall point of the cylinder head. Stall is defined as the point at which either the head gains no flow, or flow actually drops when the valve is opened more. Keep in mind that there are considerations regarding valve lift other than flow. Check for coil bind of the valve spring, retainer-to-guide clearance and, of course, piston-to-valve clearance.

If you do not want to bother with this, purchase a direct-replacement cam with approximately 0.470 inch lift, and in most instances the OE valve springs will work fine. With a stock cylinder head, there is only about a 10-horsepower potential in airflow jumping to a lift of 0.500 inch with a cam that has the same profile.

Since airflow increases greatly as the intake valve moves from its seat and farther into the combustion chamber, it is beneficial to have an aggressive intake ramp (the rate at which the valve opens). For example, at approximately 0.100 inch lift, the 1989 cylinder head flowed approximately 60.68cfm, but at 0.200 inch, it jumped to 117.78cfm, almost double.

LSA

By retaining a lobe separation angle of 114 degrees, a good idle will be achieved along with a minimum of hydrocarbon emissions.

I cannot stress enough that long-runner intake manifolds are very unforgiving with regard to camshaft duration. Runner length, not fuel delivery type, is the determining factor when choosing a camshaft. If in doubt, always choose less, not more. You will be all smiles the first time you step on the gas.

Intake Manifold and Throttle Body Choices

The Super Ram

ACCEL produces intake manifolds for both small- and big-block Chevrolet engines. Under the Super Ram series, a high-flow intake manifold and shorter runners are offered either separately or in a kit. The lower or base manifold (part number 74197) increases manifold port volume and recontours port shape for increased inlet airflow while maintaining high flow velocity and providing maximum charge density. The Super Ram Manifold Base accepts OEM TPI runners, ACCEL High-Flow TPI runners or the ACCEL Super Ram plenum/runner kit. The ACCEL Super Ram base is equipped with EGR ports for use with vehicles originally equipped with EGR.

The ACCEL Super Ram plenum and runner kit (part number 74196) are offered as a complete kit for TPI-equipped engines. Consisting of a unique box-shaped plenum and

Fuel rail and line connections and adapters are all offered by ACCEL.

This Pontiac 455 intake manifold was modified with a sheet metal plenum and a stock TPI throttle body.

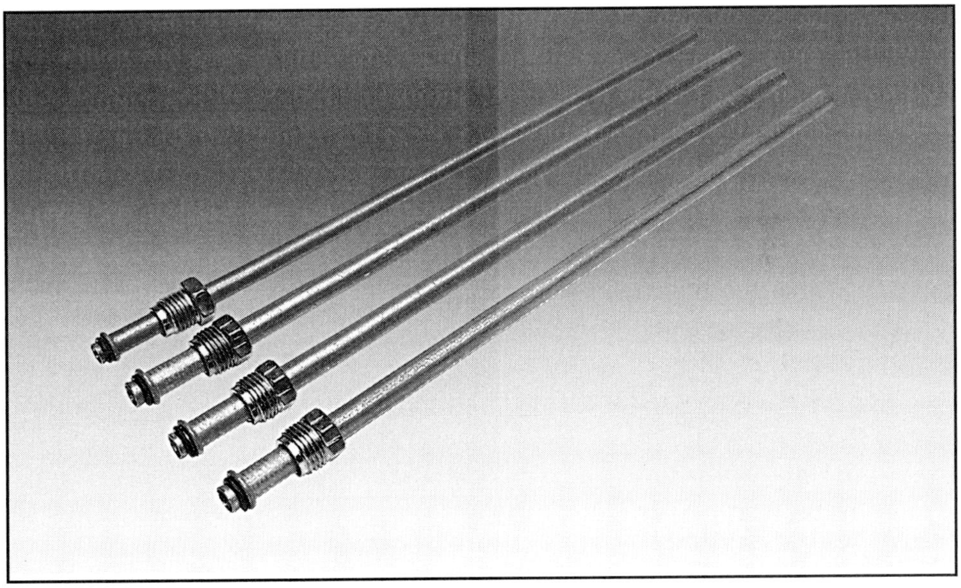

Fuel rail jumper kits allow the use of Saginaw fittings on the fuel rail and straight or NPT lines elsewhere.

approximately four-inch-shorter runners, it attaches to either a stock GM or the ACCEL 74197 base. This plenum and runner design significantly improves volumetric efficiency for increases in horsepower and torque. A separate part number (74196F) is designed for LT1-powered cars and includes a unique base manifold. This conversion is an excellent choice for automatic-equipped LT1 vehicles, and works especially well on the Impala SS.

High-Flow TPI Kit

If a stock-looking TPI is what you want but your engine demands increased airflow, then the ACCEL High-Flow TPI runner kit is for you. ACCEL engineers spent hundreds of hours on a the flow bench and at the dyno developing the optimum balance of port volume and velocity to ensure the their runner design outperforms any other existing style. These new castings are precision made and machined for unequalled port shape consistency and easy bolt-on installation. Part number 74195 fits the 1985 to 1988 TPI, and part number 74199 is for 1989 to 1992 TPIs. The earlier castings had a provision for a cold-start injector that was eliminated for the 1989 model year.

A recent introduction to the small-block family of intake manifolds is the single-plane EFI design that can be used with either a carburetor or electronic fuel injection. Part number 74140 is the manifold only; 74139 includes the fuel rails. Designed to work well with long-duration camshafts, this intake produces excellent power from 2000 to 6500 rpm and has optimized runners for precise air delivery and correct air

A Weiand Mopar Cross-Ram EFI conversion is tuned on the dyno by Rance Baxter.

Talk about runner length: This Long Ram Chrysler 440 was also built by Rance Baxter.

distribution. It was engineered to fit standard-production 23-degree small-block Chevy cylinder heads, as well as LT1 and Vortec L31 castings.

To supply air to the single-plane intake manifold, a 1200-cfm four-barrel air door is offered as part number 74202L with IAC and TPS, or 74202M without these components. Manufactured from billet aluminum and machined to close tolerances, the throttle blades are hand-lapped to provide minimum air leakage while offering excellent idle control. Stainless throttle shafts provide excellent corrosion resistance for marine use. A staged, progressive linkage guarantees unparalleled driveability.

Servicing the needs of stock-style TPI systems is part number 74190, which is a 1000-cfm billet-aluminum throttle body. It includes two 58mm throttle plates (stock are 48mm) and is CNC-machined for precise, leak-proof throttle operation. Stock GM throttle blades screw onto a flat surface machined onto the throttle shaft. High airflow entering the throttle body can exert such force as to produce bending and misalignment of the throttle blades. ACCEL solved this problem by using a slotted throttle shaft. As the throttle blades fit into the shaft's slots, clamping force is significantly increased and bending is eliminated. ACCEL also spins the end of the throttle shaft over the throttle linkage for increased durability.

Throttle Body

The ACCEL High-Flow billet throttle body fits OEM TPI manifolds as well as all ACCEL Super Ram manifolds. If ACCEL conversion kit part number 74801 is used, this throttle body can attach to any carbureted intake manifold with a Holley-style bolt pattern to convert to EFI.

The selection of the properly designed intake manifold is paramount to the success of your project, along with how calibration-friendly the engine is. Many factors need to be considered before you buy any intake manifold. If emission compliance is important, then only parts that contain E.O. (executive order) numbers can be used. Long-runner intake designs such as the stock TPI or the Super Ram, even after extensive work, are still flow-limited due to their length and poor entry angle into the cylinder head. For this reason, they are not good choices for an engine with a very good-flowing cylinder head or most racing applications. The ACCEL single-

Saginaw fittings are identified by an O-ring

plane is an excellent choice for these uses and offers a tuning length and an entry angle that work well with high-performance cylinder heads. In addition, the single-plane interfaces well with aggressive camshaft profiles that would work well with a carburetor.

Since ACCEL offers a limited number of small-block intake manifolds, you may need to fabricate your own EFI induction system. Though many have done this with sheet metal, most applications would be well-suited to converting a carburetor manifold to port fuel injection. There are so many excellent single-plane intake manifolds on the market for almost every make domestic engine that there is no use reinventing the wheel. ACCEL offers aluminum injector bushings under part number 74740 (a package of eight). To mount the bushing, simply drill or machine for acceptance, and either weld or epoxy in place. The portion of the bushing that extends into the manifold port is then ground away to maintain unobstructed airflow.

With some basic machining and knowledge of engine tuning, the benefits of EFI can be achieved on any engine.

Throttle Body Flow vs. Carburetor Size

For those of us who grew up with carburetors, throttle body sizing is often compared to wet-flow system cfm. As mentioned elsewhere, the thinking in this hobby is, the larger the part, the better it will perform. When applied to carburetors, choosing the proper flow rating is the key to obtaining the best performance. However, there has been much confusion over flow specifications for both EFI and carburetors, test procedures, and their meaning. What is an enthusiast to do? The only logical approach is to establish a full understanding of airflow testing and an engine's requirements.

The industry standard for quantifying the amount of air passing through the induction system of an engine is a flow bench test. Carburetors are identified with flow ratings to qualify their size, but cylinder head flow testing is what has made airflow numbers a common topic during bench race sessions. The problem lies in how this information is derived, and that cylinder head airflow cannot be correlated to carburetor flow numbers.

The fact has been established that one of the keys to power in an engine is airflow; so measuring airflow, in theory, should enable one to establish potential output.

Most of the confusion stems from the fact that the industry as a whole has never really established a common guideline for a flow test. It is assumed that all carburetor manufacturers use the same amount of depression as a test value, but that is not always the case. A very basic law of physics is overlooked: Two things cannot occupy the same place at the same time. Traditional flow benches are dry flow, moving air only, whereas a carburetor on an engine is moving both fuel and air simultaneously. So using a dry-flow test to determine the actual flow of a carburetor is a faulted procedure.

Even though a specific test procedure and depression are not common in the industry, it would be safe to assign a depression of 1.5 inches of mercury (in./Hg) to a four-barrel, and 3.0 in./Hg for a two-barrel. Since a flow bench is a series of vacuum motors, the higher the depression, the more powerful the bench needs to be. Carburetor manufacturers such as Holley, Edelbrock, and Carter, along with the aftermarket carburetor tuning companies, use dry-flow testing on a flow bench that is calibrated to read in inches of water, not mercury. A rule for conversion is: .5 in./Hg is equal to 6.79 inches of water. Thus, 1.5 in./Hg would be equivalent to 20.40 inches of water. As a side note, the aftermarket cylinder head industry uses 28 inches of water as a standard depression.

What is interesting about flow numbers is that they are meaningless unless you know the depression, much like a dyno sheet with corrected values to an unknown standard. With a flow bench, if you want to see higher cfm readings, just increase the depression. For example, a test value of 250cfm at 28 inches of water would equal 282.5cfm at 36 inches of water.

Confusing matters more, a race engine will produce a depression of only six inches of water at WOT. Testing at depressions higher than that becomes moot as reference numbers, but is required to try to simulate the conditions the carburetor will see on the engine. Another issue is whether all four barrels are open when testing or flowed individually and added together. This procedure is used when the flow equipment does not have the capacity to create sufficient depression for a meaningful test pressure.

With this established, the next step is to determine the required flow capacity of the engine to choose the carburetor cfm rating. For years, a very fundamental equation has been used that makes many assumptions. It ignores compression ratio, bore and stroke ratio, combustion chamber design, etc., and historically yields a required carburetor size that is a good percentage higher than the engine needs to produce the best performance. Mathematically, it looks only at maximum engine speed along with displacement. It is: cfm @ 100% volumetric efficiency = displacement x RPM/3,456.

For example, let's take a 400-cubic-inch engine that will spin to a maximum of 6,000 rpm. In theory, using the above equation, the engine would require a 694.44-cfm carburetor. Now let's explore other considerations. Assuming that at peak torque this engine would reach only 85 percent VE, the required airflow now becomes 590.27cfm. If all aspects of the engine remained the same but the maximum engine speed was altered to 4,000 and 7,000 rpm, a carburetor of 464.96cfm and 810.18cfm at 100 percent VE would supposedly fit the bill, respectively.

Another way to examine the requirements is to consider the cfm required at 20.40 inches of water to support the horsepower of the engine: cfm = 1.68 x HP. Inserting numbers, a 400-horsepower engine would then require a 672-cfm carburetor.

With EFI, the throttle body needs to flow only air, and thus is not required to cause a depression and initiate fuel flow from a booster. Due to this, the engine becomes less sensitive to the size of the throttle plates, when kept within reason. As with the carburetor industry, EFI companies have not revealed the depression used to determine airflow. According to my sources, ACCEL has rated all of its components for flow at 28 inches of water, a higher depression than the 20.4 inches that the wet-flow people use. The age-old EFI question is, can you over-throttle body an engine? The answer is yes and no. As the throttle plates grow larger, there will be a greater throttle opening at any given throttle angle. This can now create a need for additional throttle-based acceleration enrichment programming to compensate for the greater airflow. In addition, tip-in drivability is often disturbed when an excessively large throttle body is installed.

So as much as things change, they remain the same; bigger is not always better. A 58mm throttle body on a mild 305 will hardly yield any power gain but will make the engine very digital at low throttle openings, making smooth driving very difficult.

5

Matching the ECU to the Engine

Most of the spark/fuel ECUs look similar externally regardless of whether they are meant for Buick, Ford, or Chevy, port or throttle body injection, OE-style ignition or inductive pickup. In no instances are these interchangeable; all have major internal differences for ignition compatibility. In the case of the throttle body ECU, a major design alteration is required for the firing sequence of the injectors. The simultaneous double-fire, sequential and marine versions also occupy unique enclosures and are both visually and internally different.

The 015013 wire list pertains to all spark/fuel ECUs regardless of the vehicle designation, and only the marine and SEFI share a dedicated wiring diagram.

Ignition Compatibility

The internal changes that are made to a standard spark/fuel ECU (GM HEI) are limited to ignition compatibility. The easiest way to think of ECU designation is to identify the type of ignition being operated. Since the 015013 is coded for simultaneous double firing of the fuel injectors, the ignition firing order is moot and gives the ECU wide-based installation appeal. It is very common to see this ECU running fuel-injection systems on imports, domestics other than Ford or Chevy, and industrial engines.

To identify the ignition requirements, you first need to determine with a Ford or GM application whether a standard OE-style electronic timing control is going to be used. This type of distributor has no provision for either mechanical or vacuum advance units. If electronic control of the timing is going to be used with a Ford or Chevy ignition, the ACCEL part numbers will identify it. The difference in the Chevy ECU listings (part numbers 74021, 022, 023, 024) refer to fuel and spark calibration changes for popular cubic-inch-displacement engines.

If ECU-controlled timing is desired but there are no electronic spark timing distributors that easily interface with the ACCEL ECU (Mopar, Honda, industrial), then control of the spark can still be accomplished with the use of an inductive pickup, either in a distributor or on a crank sensor. For the ignition to work with this system, an aftermarket ignition amplifier, such as an ACCEL 300+, MSD, Mallory, etc.,

All ECU development work was done in Wixom, Michigan, under the supervision of EFI guru Joe Alameddine. Here MSD ignition components are undergoing compatibility testing with a 6.0 ECU.

In the unlikely case that your DFI system develops a problem, a complete repair and upgrade staff is located in the Cleveland facility.

Joe Alameddine performed work on a modified ECU for Ford Industrial engines using a natural-gas-fueled turbocharged Triton V-10 engine. Eventually this ECU will be installed around the world, operating backup electrical generator systems during emergency conditions. This engine will experience 10 psi of boost for weeks at a time.

would be required. The ACCEL part number 74042 would need to be ordered, which is a standard ECU with a board modification to accept the inductive pickup signal. If you already have a standard spark/fuel ECU and want to have it converted from an OE timing control to an IPU, it needs to be sent back to ACCEL to have a 74043-I board modification performed. Once an inductive pickup modification is done, the ECU becomes nondescript and is not identified as a Chevy or Ford ECU.

ACCEL part number 74025 is a Buick GN spark/fuel ECU and is coded differently for the recognition of the hall-effect crank and cam sensor signals. An inductive pickup generates a modified sine wave output signal, but the Buick uses a square-wave input for its ignition. The Buick ECU modification consists of an internal PROM and software change.

The Chevrolet LT1 ECU (74022-L) also consists of a board change to interface with the output of the Delco Opti-Spark distributor. If the primary triggering function is to be moved out of the Opti-Spark and an IPU crank trigger is employed, then part number

In 1992, ACCEL converted this Chevrolet pickup to TPI with DFI engine control as a durability test vehicle. It now has almost 400,000 miles and is still using the same ECU, and it has experienced no failures.

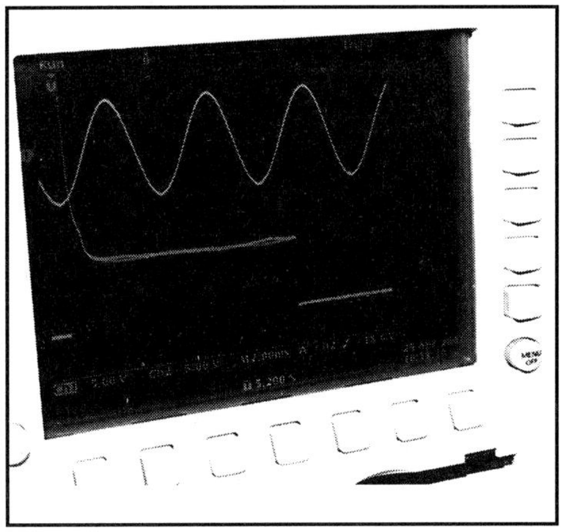

Here an ACCEL technician tests an EU for analog input and square wave output.

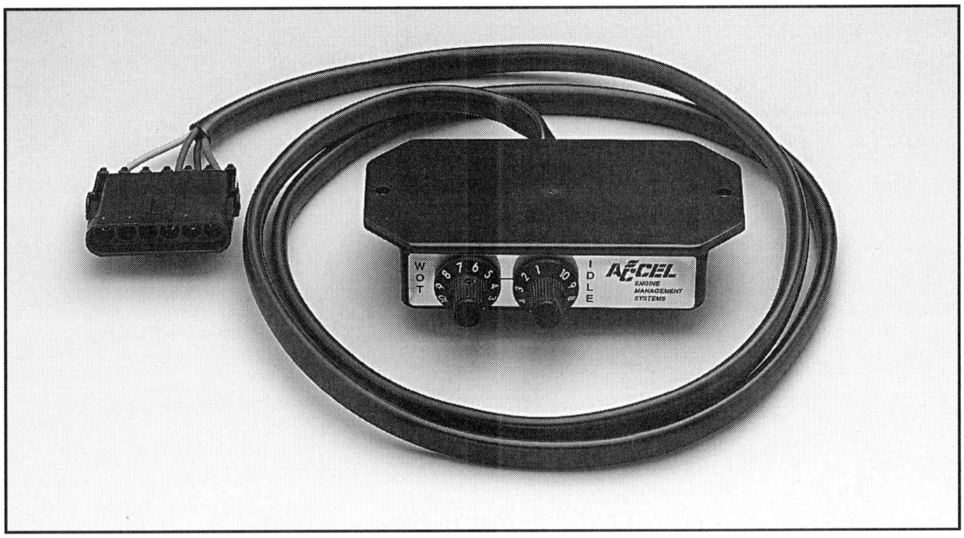

The UIM is an excellent addition for the end user who wants to do tuning at idle and WOT without CalMap.

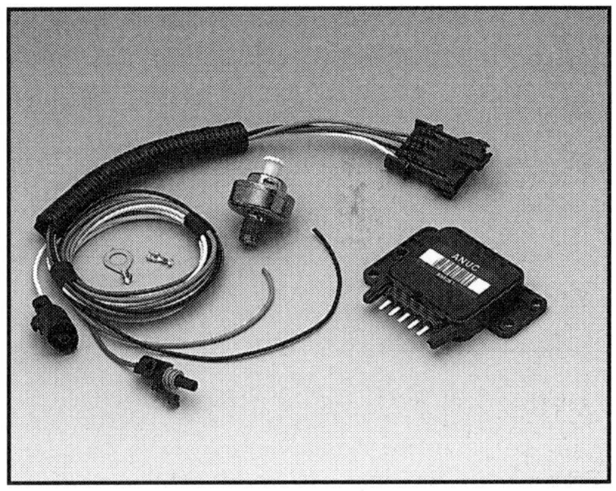

A knock control kit is an excellent addition for any nitrous or forced-induction engine.

74042 needs to be ordered.

Also offered is a marine ECU that simply consists of a 74022 in a weather-tight enclosure with water-resistant bulkhead connections. This ECU is used in boats and is also commonly found in late-model Corvettes that require the mounting of the ECU under the hood. Other uses for this part number are industrial or severe-duty environments.

Sequential spark/fuel ECUs that are offered to interface with OE electronic timing controls are applicable only to Ford Thick Film Integrated ignition and Buick GN configurations. All other SEFI ECUs are designed to work with an inductive pickup. IPU configuration is also available for Ford. Other than ignition compatibility, all SEFI ECUs have dedicated injector wiring harnesses for proper phasing of the fuel injectors.

ECUs for ACCEL throttle body kits are coded to fire the injectors differently from a conventional ECU. If electronic timing control is desired for a throttle body unit and an electronic spark timing distributor is not offered, then an IPU modification must be installed.

Both the standard spark/fuel and the SEFI ECUs have a number of options available to enhance their operation. All options that are offered can be installed at the time of purchase or added later as the needs of the engine-management system change. The following are actual hardware options, not changes in configuration through the CalMap software.

As mentioned previously, a very popular option is part number 74043-I, which allows the use of an inductive pickup distributor or crank trigger as the primary switching signal. The benefit of a crank trigger is the accuracy in identifying the piston position in the bore from the crankshaft. This is in lieu of the distributor, with its inherent stack-up

015013	1 Bank Injectors, IAC, Converter, Fan, 1 Step NOS
015014	1 Bank Injectors, IAC, Fan, Linear 02
015015	1 Bank Injectors, IAC, Converter, Linear 02
015016	1 Bank Injectors, IAC, Linear 02, 1 Step NOS
015017	1 Bank Injectors, IAC, 3 Step NOS, Fan
015018	1 Bank Injectors, IAC, 3 Step NOS, Converter
015019	1 Bank Injectors, 3 Step NOS, Linear 02, Fan
015020	2 Bank Injectors, IAC, Converter, Fan
015021	2 Bank Injectors, IAC, Linear 02
015022	2 Bank Injectors, 3 Step NOS, Converter, Fan

Use this chart to identify an unknown ECU.

Ignition distributor with induction-type pulse generator
Top: Principle. Below: Induced voltage.
1 Permanent magnet, 2 Inductive winding with core, 3 Variable air gap, 4 Rotor.

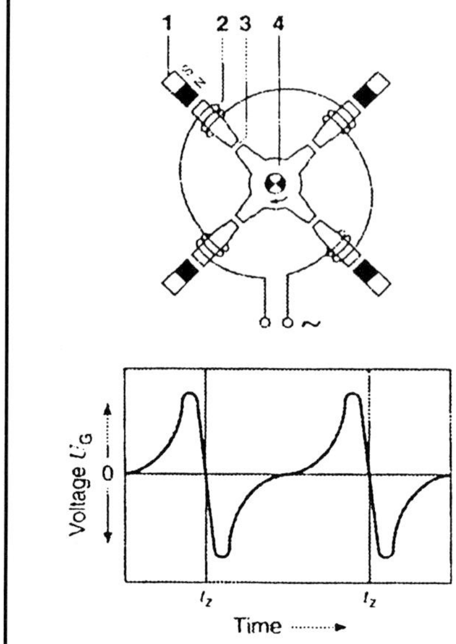

An IPU ECU reads a sawtooth sine wave from either a distributor or a crankshaft trigger for an rpm signal.

Keep the Voltage Up

The E in EFI means that the proper voltage is essential for the DFI system to function. Often a very elusive problem with a calibration can be traced back to low voltage. A 6.0 ECU will function best with 14.6 volts or greater as an input. Once the voltage drops below this threshold, a drivability issue may occur. If the alternator output goes below 12.5 volts, then idle surge, roll, and intermittent stalling conditions are almost guaranteed. The 6.0 ECU has no internal voltage correction table, so as the power input drops, the injectors open more slowly and create a lean condition. Factory ECUs have correction factors that allow acceptable performance down to approximately 9.6 volts, making them more forgiving.

The most common cause of low voltage, especially at idle, is the use of improperly sized underdrive pulleys. They slow down the alternator to such an extent that it can barely produce enough voltage to keep the engine running. Underdrive pulley manufacturers such as March offer many different pulley sizes, so the benefit of reduced drag can be realized without inducing a charging circuit problem.

The voltage output of an alternator can be confirmed by connecting a voltmeter across the battery terminal. It's essential to check the output at idle with a full accessory load (high-beam headlights, air conditioner, wipers, rear window defogger, and radio) on. In the early days of my tuning career, I had a turbocharged 5.0 Mustang that ran great until I drove it home at night. When I got to the second traffic light, the voltage degraded enough that the engine would surge and stall. During the day it ran fine. The problem was low alternator output from improperly matched underdrive pulleys.

of tolerance from the timing chain and interface with the distributor gear. This option also allows the use of the ECU-controlled programmable timing on engines that did not already incorporate an electronic spark timing distributor, or were not originally ignition-compatible with the ACCEL ECU.

User Interface Module

The acronym UIM is used to describe a stand-alone interface that gives the end user the ability to alter the air/fuel ratio that is programmed into the core fuel table, either richer or leaner at idle and wide-open throttle. The module incorporates two knobs: one for idle, the other for WOT. Simple to install, the four leads plug into ECU bulkhead connector P1 at terminals C6, D4, C4, and D8. The UIM has no control over the mixture

The engine from the ACCEL pickup has been subjected to all types of driving to test the ECU under real-life conditions.

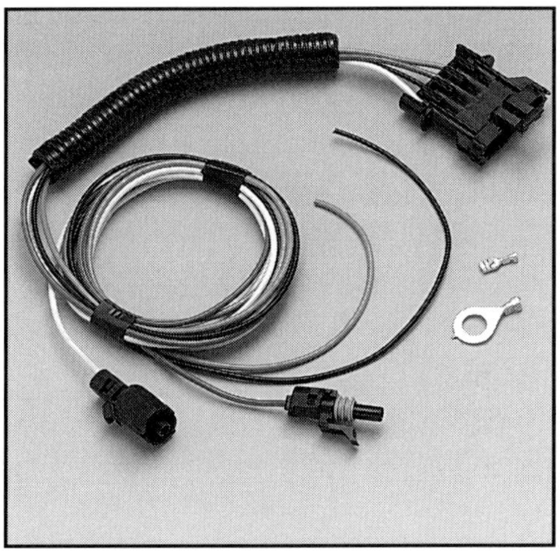

A knock sensor wiring kit can be purchased if you already own the module and sensor.

above one percent TPS and below 99 percent TPS. This is an excellent option for the person who likes to do some minor tuning but is not interested in mastering the CalMap software.

Linear Oxygen Sensor

The linear oxygen sensor option requires a PROM change to the ECU. This kit includes a NTK-sourced wide-band controller (known as a brick), NTK 5-wire oxygen sensor that replaces the standard issue Bosch unit, and the corresponding wiring harness to attach the above to the main wiring harness. The linear feedback option gives the user the ability to have multiple target air/fuel ratios for total control and correction over the mixture.

The linear oxygen sensor system works in contrast to the standard ECU, which, if operated in closed loop, has a target air/fuel ratio of 14.7:1 as its only set point. The standard lambda = 1 design is programmed to correct to that target mixture only under certain operating conditions. The linear option allows target air/fuel ratios referenced from every two rows of the fuel matrix based on load, ultimately providing eight target air/fuel ratios. This option also provides the ability to set a target warm-up air/fuel ratio that will be

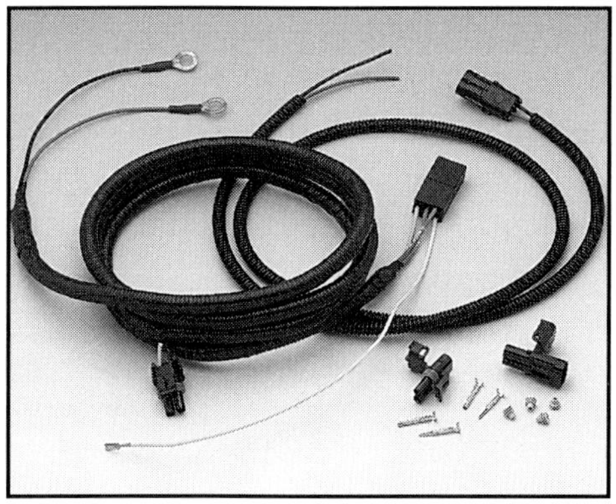

A TCC lockup kit is offered for GM transmissions.

maintained after the first 45 seconds of startup and until the closed-loop temperature parameter is met in the CalMap software. Deceleration enleanment through fuel shut-off monitors TPS and RPM and offers programmable set points to shut down the injector pulse width during closed-throttle conditions for better coast-down emissions control.

Altitude Adjustment

ACCEL offers an on-the-board change to the ECU and an additional one-bar MAP sensor that is referenced to atmosphere to function as a trim to the base fuel map for weather and altitude changes. This option uses the same ports in the bulkhead connector as the UIM and cannot be used in conjunction with the UIM.

Three-Stage Nitrous

The standard spark/fuel ECU has the ability to control one stage of nitrous, with the optional ability of adding two additional stages. To add this control, ports P2-B5 and B4, which originally controlled the coolant fan and TCC functions, must be lost.

Interface Harness

Sources other than ACCEL produce a jumper harness that allows the use

Fast Track Performance (888-640-3278) makes any jumper harness to integrate a DFI into almost any engine-management system.

of an ACCEL ECU through the OE wiring harness. This eliminates the need to remove the factory wiring harness and makes the installation of a DFI a half-hour job. The benefit of this style of installation is the ease of connection without cutting the car's original harness, along with the ability to revert to stock engine-management electronics if so desired. A downside of this installation is that in many cases the jumper harness is wired not to use all of the available drivers in the ECU, and this, in turn, limits the range of fuel injectors that can be operated.

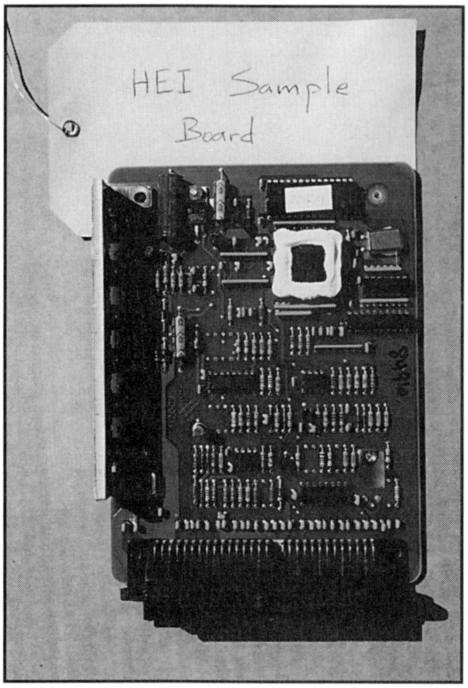

GM HEI EST circuit board.

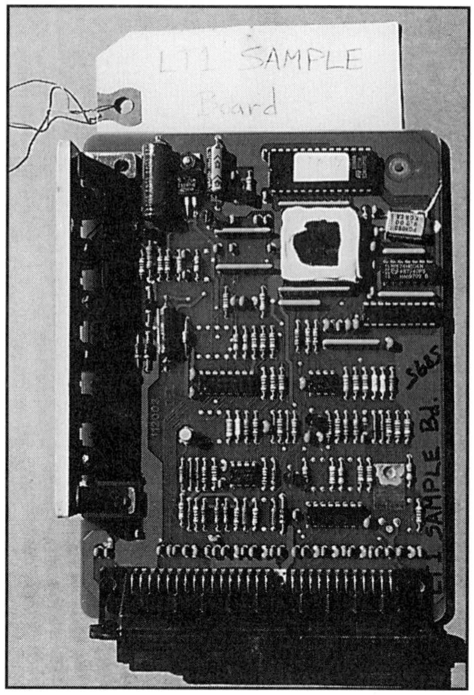

LTI ECU circuit board.

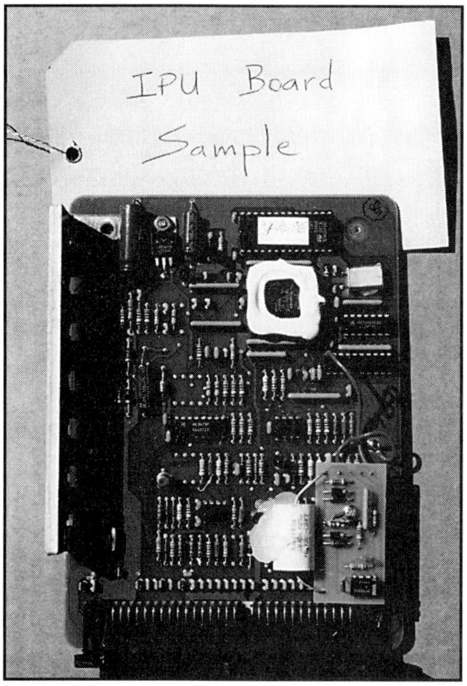

IPU circuit board.

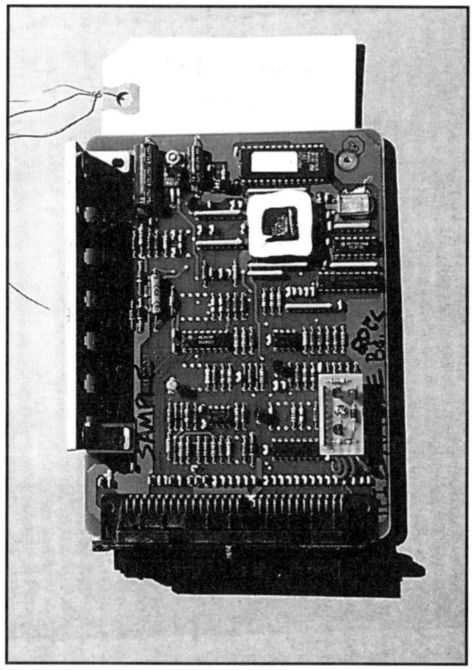

Ford TFI circuit board.

Getting to Know VIC

Anytime that tuneability is added to a fuel-injection system, the reward comes back many times over. Programmable ECUs offer adjustability that is not available with an OE unit. Occasionally, even a DFI system leaves you hoping for additional tweaking capabilities in your quest for maximum performance. Thanks to the efforts of Dan Kapers and Joe Alameddine, who are both development engineers at ACCEL, a new level of control has been developed. It is the Variable Injector Controller or VIC (part number 74070).

It is designed to work in conjunction with the following batch-fire (non-sequential) systems: ACCEL/DFI, GM ECM 1991 and earlier, Ford ECU 1990 and earlier, Haltech, MoTech.

The VIC was engineered to offer the ability to alter the injector pulse width by percentage on the fly, on a per-cylinder basis, without the use of any software. This is accomplished by simply turning the appropriate knob on the unit. VIC has the ability to richen (longer injector pulse widths) up to 25 percent. ACCEL was not content with bringing to market the only product that allowed individual cylinder mixture control on a batch-fire system, so it also included the ability to operate eight low-impedance injectors on any of the above-mentioned engine controllers. No longer are the drivers in the ECU an obstacle to integrating a low-impedance, high-flow injector. With the installation of the VIC, the injector control circuitry is now moved from the ECU.

The VIC is manufactured with a black anodized extruded aluminum case that incorporates integral heat sinks along with eight individual cylinder-tuning knobs. Each knob is clearly identified for both cylinder control and the percentage of injector pulse width modification. As a safety feature, to prevent accidental changes to the fuel curve, an Allen-head lock screw secures positioning. The VIC also includes a simple-to-understand installation manual along with an OEM-quality injector and power harness with an integral relay and fuse.

A breakthrough in technology allows the VIC to employ switching circuits instead of drivers to open the injectors. The efficiency of this circuit actually created a slight obstacle in the early development stages, since the VIC increased injector response time by 65 percent. ACCEL then worked with the design to trim that value down to 40 percent to avoid a rich condition and recalibration of the fuel map after installation. Joe Alameddine stated, "We wanted to assure this product was as user-friendly as possible, so we cut the increased rise time down to 40 percent." He added, "The VIC is able to offer repeatable short pulse width duty cycles down to a charging circuit voltage of 9.6 volts DC."

The electronics in the VIC are very accurate and stable, boasting a maximum deviation of less than one percent at 25 percent correction. Testing by ACCEL has shown that the average engine needs to be tweaked only three to eight percent between the leanest and richest cylinders.

6

An Introduction to CalMap Software

Before any programming can be accomplished, the CalMap calibration software needs to be understood and mastered. Since this is a multi-dimensional process, this chapter, along with Chapters 7 and 8, will be used for this purpose. In this part of the book, the mechanics of CalMap will be discussed, consisting of PC requirements, hardware installation, and running modes. Chapter 7, "Working with the CalMap software," provides a screen-by-screen explanation, and Chapter 8 covers the actual tuning process.

Overview

CalMap provides you with full access to all of the engine management control tables in the ACCEL ECU. It also allows the monitoring of all the sensor signals that the ECU receives from the engine. A desktop or laptop personal computer uses the CalMap software to communicate with the ECU. When setting the engine management functions or calibration with CalMap, adjustments can be made both "on-line" and "offline," in the vehicle, on the bench, or on the computer disk. The tables stored in the ECU's non-volatile memory (not lost when the ignition is turned off), which control fuel injection, ignition timing, and idle speed, are fully accessible. These tables can be changed to fit almost any engine configuration. Naturally aspirated, turbocharged, and supercharged engines all can be accommodated with the ACCEL ECU and software package. Monitor-ing engine performance and system troubleshooting are easily achieved using either the data log or view functions.

Here are the minimum computer requirements for operating CalMap:

1) Laptop or desktop PC with DOS version 2.11 or greater, RS-232 serial port, and 3 1/2-inch or 5 1/4-inch disk drive. The minimum recommended CPU speed is 8 Mhz; minimum memory is 512K RAM.
2) 640x200-pixel display capability (VGA) or graphic display.
3) A 3 1/2-inch or 5 1/4-inch floppy disk containing ACCEL's CalMap software.
4) A serial communications cable from ACCEL.

Although it is preferable to perform most calibrations on a dynamometer, the system can be easily calibrated in the vehicle. When performing a

In-Car Laptop Tips

The use of a laptop in a car is, at best, very awkward. Even though it is best to tune on a chassis dyno, many driving scenarios cannot be duplicated and will force on-the-road tuning. Cold start, torque converter lockup, acceleration enrichment, IAC tracking, and stall-saver functions are best done with the vehicle on the road.

Successful tuning of any function that requires the vehicle to be driven demands a helper to drive. We have all tried to tune and drive, but it is a very dangerous proposition, even if you live in the most rural of areas. From real-world experience, it takes two to three times longer to do an on-the-road calibration by yourself than with a driver. Not being concerned with the road allows your full attention to be focused on the calibration screens and how the engine is responding.

Always use a 12-volt adapter for your computer since most calibrations take longer than you think, and a dead battery in the PC will end the session abruptly. Use a rubber band to tie up and remove unnecessary length from the power and CalMap cords. Position the PC on your lap or use a small wooden board as a rest for it on your knees. If you get motion sicknesses easily, make sure you limit the time spent looking at the screen while the vehicle is moving. Take along a calculator so that changes to the fuel map can be made mathematically by dividing the produced air/fuel ratio by the actual mixture. Alter the matrix values by the derived percentage.

Before tuning on the road, make sure the starting, idle, and WOT calibration are already done. This will lay the architecture for the rest of the tables.

calibration, both ACCEL and the author recommend that one person operate the vehicle while the passenger alters the calibration. (See above.)

How CalMap Works With PCs

CalMap is a program that runs in your PC. It acts as a translator between you and the ECU. Several memory files are also included on the CalMap disk. These memory files contain information about how the ECU memory is laid out. When CalMap first begins to communicate with the ECU, it reads this information from the appropriate memory file so that it knows where to look for data in the ECU. The Cal Map disk also has a filing system. There are about a dozen directories present on the disk, which keep tables that you save in an orderly arrangement. For example, if you save a Cranking Fuel table from the ECU on the disk it is put into the CRANK directory, the Base Fuel table goes into the BASE directory, etc.

To better understand how CalMap works, let's look at how a PC operates. A PC can have many things going on inside it simultaneously. Think of it as a corporation: Many tasks are performed by many different managers. For the company to work smoothly, all of the managers must be coordinated to perform the correct tasks at the proper times to produce a final product. Some type of structured operating environment is required to make this happen. In the PC this structure has a name. It is called DOS, for disk operating system. DOS coordinates all the filing, interfacing with customers, and quality control, as well as many other functions of the corporation.

While DOS is responsible for the overall operation of the PC, there are many individual workers in the corporation that make up the workforce. The workforce in the PC is called the BIOS, or basic input-output system. Within this group are many different individual employees. One takes care of the keyboard, one the display, one the disk drive, etc. As an example, consider the entry of a word from the keyboard. The keyboard worker gets each of the letters entered from the keyboard and gives them to the boss (DOS). The boss gives each of the letters to the screen worker, who displays them. Once the ENTER key is struck at the end of the word, the boss is informed that a customer request has been issued. The boss then plans the work that needs to be done to fulfill the customer's request and distributes the proper instructions to the various departments to execute.

When the CalMap programs runs, it tells DOS, "I'm taking control now. I'll let you do only the things I do not want to do." This arrangement works out quite well, since DOS does not know anything about engine controls, and CalMap doesn't know much about running a PC. As it turns out, DOS is very cooperative and only squawks when one of its employees is mistreated. When this happens, the PC will let you know with a sound.

Installing CalMap

When you receive CalMap, you will have the following in the package:

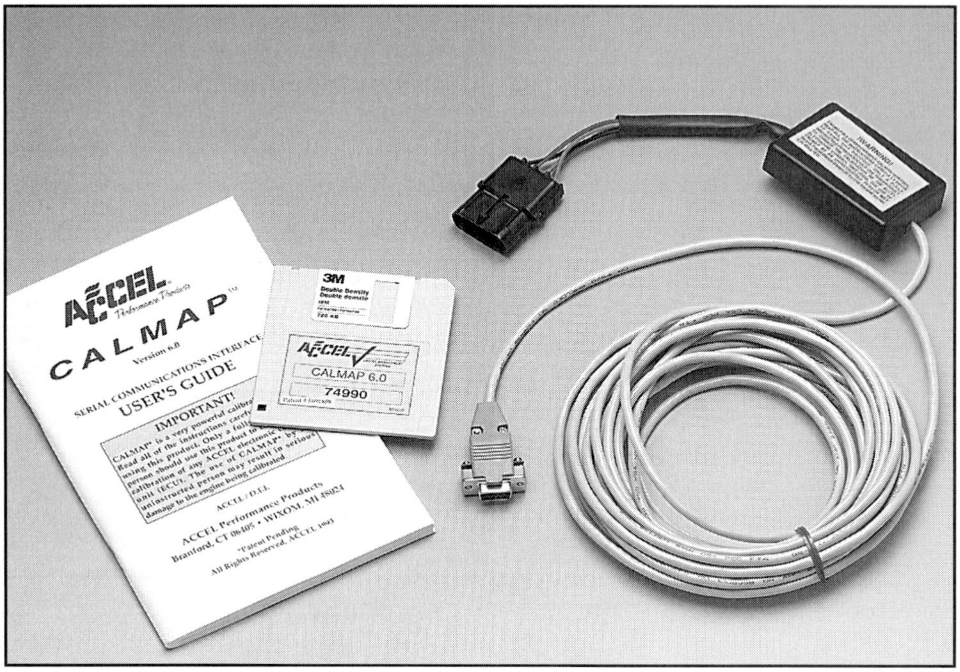

CalMap software comes with everything you need to link your PC to your car's ECU.

- one 3 1/2-inch floppy disk
- one 5 1/4-inch floppy disk
- one 5-foot or 25-foot interface cable
- one set of instructions

Connect the ACCEL-supplied serial cable to your PC and to the harness at the ECU. The cable has a small module mounted in the line. Do not damage or attempt to disassemble this module. The 9-pin connector end of the cable connects to the RS-232 serial port on the PC. If your PC has a 25-pin serial port connector, you will need to get a 25-to-9-pin adapter, available at any computer or electronics store. If your computer has more than one serial port, you must use the Com 1 port.

To begin using CalMap, follow these step-by-step instructions for your particular PC configuration.

One-Disk-Drive Computer

Step 1: Insert your DOS system disk in the drive.
Step 2: Turn on the computer.
Step 3: When the DOS prompt appears on the screen, remove the DOS system disk.
Step 4: Insert the CalMap disk in the drive
Step 5: Type CALMAP and press ENTER.

Two-Disk-Drive Computer

Step 1: Insert your DOS system disk in Drive A. Insert the CalMap disk in Drive B.
Step 2: Turn on the computer.
Step 3: When the DOS prompt appears on the screen, type B: and press ENTER.
Step 4: Type CALMAP and press ENTER.

Hard Disk

CalMap must be installed on the hard disk by copying the entire contents of the CalMap disk to your computer's hard disk before running.
Step 1: After the DOS prompt appears, create a new CalMap directory on the hard drive by typing "md CALMAP." Then press RETURN.
Step 2: Insert the CalMap disk in drive A.
Step 3: Change to the CalMap directory by typing "cdCALMAP." Press RETURN.
Step 4: Copy the files from the CalMap disk to the CalMap directory on your hard drive by typing "copy a.*.*." Press RETURN.
Step 5: Type CALMAP and press RETURN

This screen will appear when CalMap is booted up on your PC.

Once CalMap begins running, the screen will display the ACCEL logo. Below the logo, two function selections will appear: Edit ECU and Edit File. A selection box will surround one of the functions. The selection box can be moved using the arrow keys. When you press the ENTER key, the surrounded function will begin. Press the Q key to return to DOS.

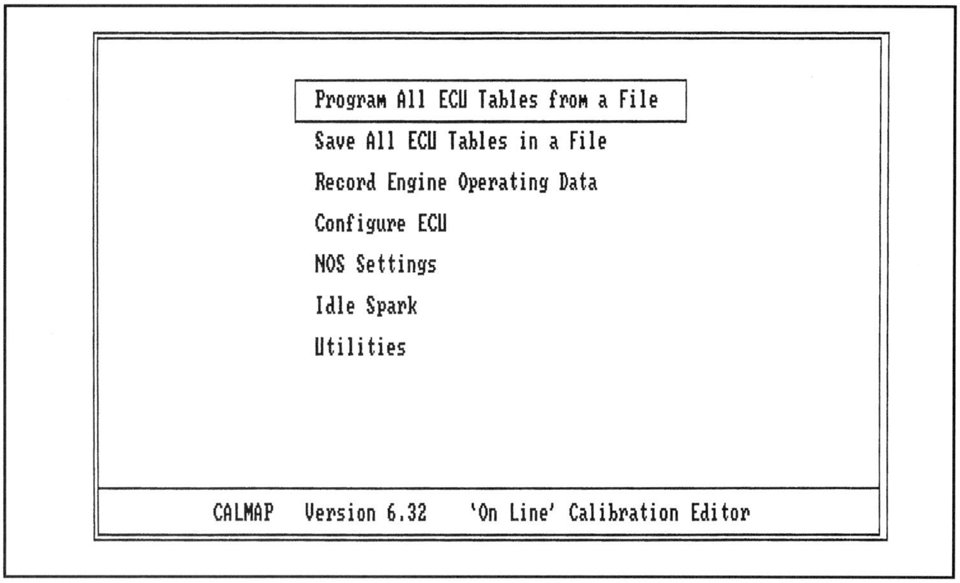

The Global menu offers many functions ancillary to the ECU's operation.

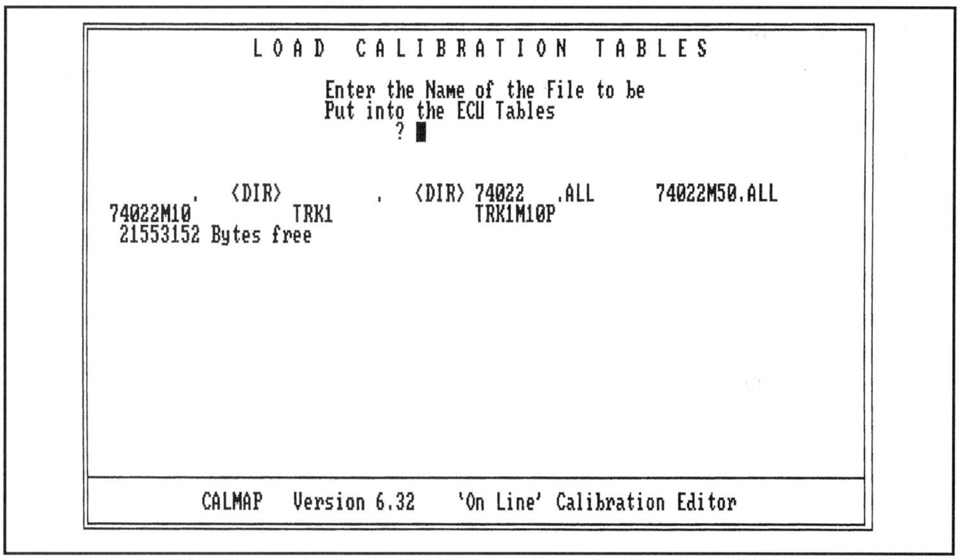

It is advisable to save the complete calibration in a Global file.

Getting Started

The following is a brief overview of the various menus and commands of CalMap.

Edit ECU/Edit File

The *Edit File* function allows you to look at and change tables that were previously saved from an ACCEL ECU. No connection to the ECU is required. The *Edit ECU* function lets you communicate directly with the ECU. Fuel tables, ignition tables, and other engine operating data are imported to you via the serial communication cable from the ECU and displayed on the screen. As long as power is supplied to the ECU, you can communicate with it. Communication with either the ECU, using CalMap, can be established whether the vehicle's engine is running or not. The ignition key must be in the "on" position, but the engine does not have to be running. The ECU can be reprogrammed while the engine is running.

Once the *Edit ECU* function is selected, a car will drive across the screen and display the message "Waiting for ECU response" if it does not get a response. This prompt will continue indefinitely until communication is established. If the PC tells you there is no response from the ECU, make sure that the serial communications cable is connected properly.

Once contact with the ECU is established, the screen will change from the ACCEL logo to a root menu. By pressing the up/down arrow keys, a selection box is moved from selection to selection, similar to looking over a menu. To choose any one of the submenus on the screen, move the box to a position where it is surrounding the desired submenu and press ENTER. To exit the menu and return to the point where you started, press Q.

Everything in CalMap is accessed from the root menu. After a selection is made in the root menu, a submenu will take its place on the screen. From each submenu, a specific function can be selected. The mechanisms that allow you to move through CalMap are the arrow keys, the ENTER key, and the Q key.

The ENTER key selects the

function that you are pointing to and serves to move you forward in the CalMap program.

The Q key moves you backward in the menu structure so you may exit from any of the submenus or CalMap itself.

Save ECU Tables in a File and Program ECU Tables from a File—For now, do not dwell on what each selection allows the engine calibration to do; that information will be covered in the next chapter. Go to some of the submenus from the root menu by moving the selection box with the arrow keys and press ENTER. Pressing Q will return you to the root menu. As you go from submenu to submenu, you will soon see that each submenu has two selections in common: *Save the ECU Table in a File* and *Program the ECU Table from a File*. These commands, when selected, do exactly what their labels imply.

Save the ECU Table in a File first reads the appropriate table from the ECU and then saves it on your disk under a name that you assign. *Program the ECU Table from a File* puts a table you have previously saved into your ECU.

When *Save the ECU Table in a File* is selected, CalMap will read the entire table, or group of tables, associated with the particular function. For larger tables, a bar graph will appear on the screen to notify you of the progress of the reading process. Once the table data is read from the ECU, the screen will list all the previously saved table files. Above the file list, a question will appear asking you to name the new file. Type in the name you wish to assign to this new file. Unfortunately, there are limitations when naming files. File

```
Base Fuel Map
Cranking Fuel
Idle Speed
After Start Enrichment
Warm-Up Enrichment
Air Temperature Correction
Acceleration Enrichment
Ignition
Global

CALMAP   Version 6.32   'On Line' Calibration Editor
```

The main menu is where most of your tuning work will be done.

names can contain only letters and numbers (alphanumeric character), and there can be no more than eight characters in the name. However, file names can also have up to three character extensions. Some examples of file names are: 305280, 350.IDL, 383TURBO.MAP, etc., where the extension is the set of characters following the dot (period). Once you have typed the new file name, press the ENTER key to save it.

After the file has been saved, you will be automatically returned to the corresponding submenu. If, after the table has been read from the ECU, and the file list has been displayed, you decide that you do not want to save the table after all, do not enter a new file name. Just press the ENTER key and you will be returned to the submenu without having saved the table data in a file.

Note that for each submenu there is a corresponding directory on your disk in which CalMap saves the corresponding table data. The idle control tables are saved in the IDLE directory, the air temperature correction tables are saved in the AIR directory, etc. Also note that before you enter a name for the file in which to save a table, CalMap lists all of the files in the present directory so you can determine if you will be writing over an existing file with the same name. Sometimes writing over a file is desirable in order to minimize confusion when repeated changes are made to the same table during a calibration. At the same time, it can be a good idea to save past table files. Then they can be retrieved if experimentation during an engine calibration has brought you to an unacceptable point. The method you choose is strictly a matter of personal preference.

When *Program the ECU Table from a File* is selected, you may program new table values into the ECU from a file that has been previously saved by CalMap. First, a list of the files in the directory associated with the submenu you are in will appear on the screen. These files are your only choices for programming the ECU at this time. Once you have created more files, the

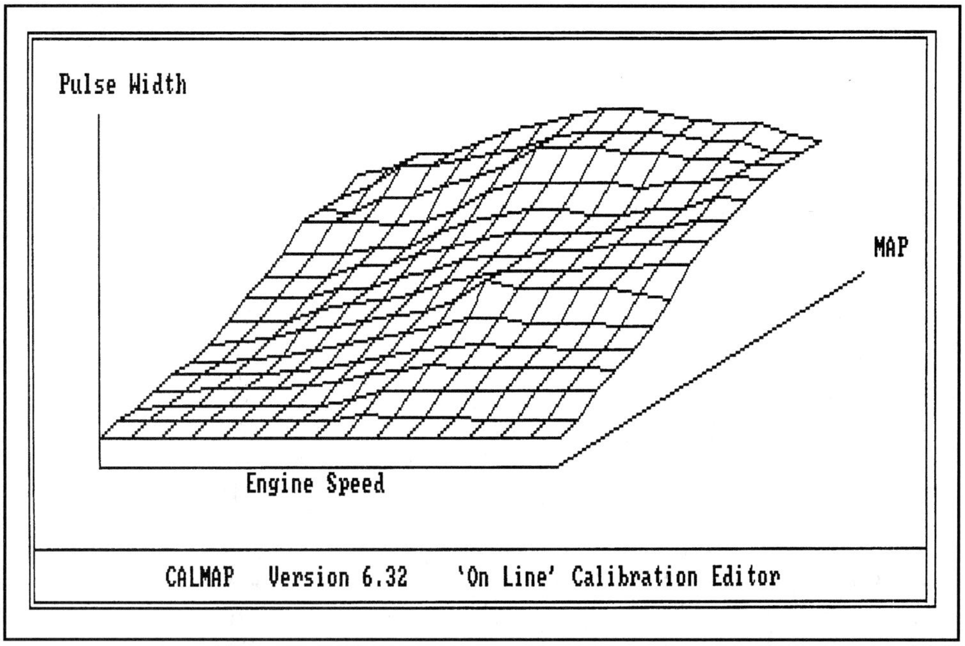

By graphing the Base Fuel Map function, any errors can be easily seen. This is an example of a nice, smooth fuel curve.

number of choices from the directory will grow.

After the files have been listed, you will be asked for the name of the file you wish to program into the ECU. If you decide not to program a new file into the ECU, just press the ENTER key without typing in the name of a file and you will be returned to the submenu from which you started. To program a table into the ECU from a file, type in the name of the table file you wish to put into the ECU and then press ENTER. CalMap will begin programming your ECU from that file and then return you to the submenu from which you started. If CalMap has a problem finding the file you have requested, it will honk and display an error message. Press Q and you will be returned to the submenu for another try.

The third and/or fourth entry in each of the submenus is where it all happens. Viewing or editing the actual function occurs here. The display will be either a graph or a matrix of numbers. In all cases, you are allowed to reprogram the actual curves in real time (during operation of the ECU).

Base Fuel Map Submenus

After *Base Fuel Map* is selected in the root menu, the root menu will disappear and a submenu will list four choices. The first two choices, if selected, function as described previously. When you enter *Program Base Fuel Map from a File*, a matrix of 240 numbers is downloaded to the ECU from a file that has been saved on a disk. When you enter *Save Base Fuel Map in a File*, this matrix of numbers is saved from the ECU to a file on the computer disk. The matrix numbers make up the *Base Fuel Map* for the ECU and determine the fuel flow at each possible speed/load operating point of the engine. The fuel map consists of 15 rows and 16 columns of numbers. The rows correspond to the manifold absolute pressure (MAP), and the columns represent the engine speed (rpm).

The lower left portion of the screen will contain the injector pulse width. This is the time that the injector is opened in order to allow fuel to flow through the nozzle. A millisecond is 1/1000 of a second. The pulse width is labeled *Base Pulse Width.* While in the Base Fuel Map, CalMap continuously receives data from the ECU and updates the screen with the new data as it is actually changing during engine operation. The location of the flashing cursor box in the Base Fuel Map indicates the cell in which the ECU is currently looking for an injector pulse width. The cursor correlates to load, manifold absolute pressure (MAP), and engine speed, based on the sensor inputs. If the engine is not running, the cursor will remain in the lower left corner. During engine operation, as the engine speed increases, the cursor will mover to the right. As manifold pressure increases, it will move downward.

Graph Base Fuel Map—When *Graph Base Fuel Map* is selected, the fuel program is uploaded from the ECU to the CalMap disk in the PC. A bar graph on the screen showing the progress of the upload will indicate when the upload has finished. After all of the data is received by the PC, a three-dimensional plot of the fuel matrix is generated and displayed. The point on the plot where the vertical axis meets the horizontal axis is where the injector pulse width and engine speed are both 0. As you move up the vertical axis, the pulse width increases. As you move to the right on the horizontal axis, the engine speed increases. The point where the horizontal axis meets the axis going back into the screen is where the engine speed and the manifold

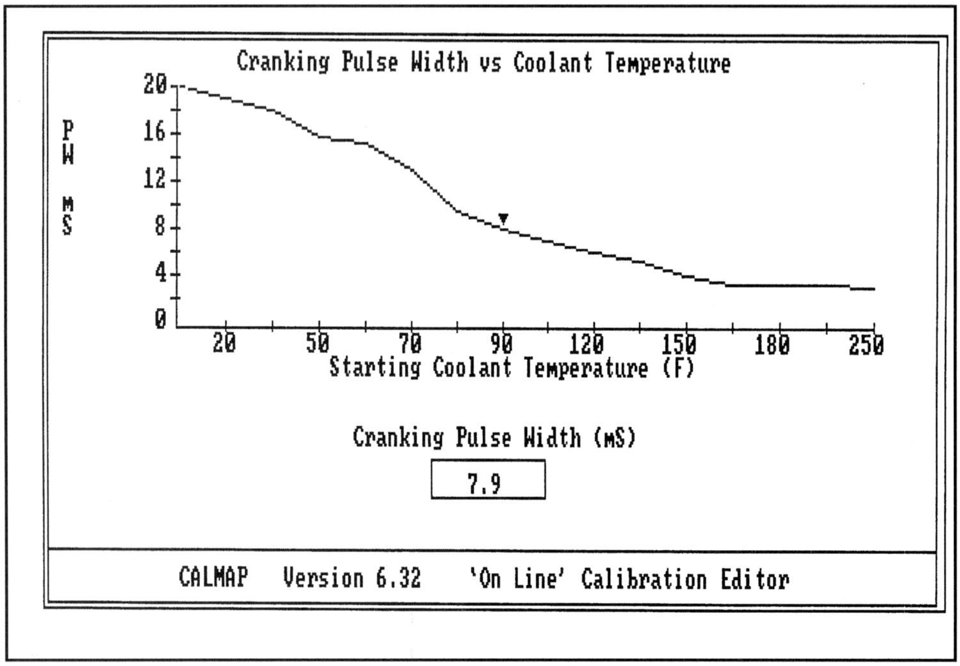

Cranking fuel works independent of the values in the fuel map.

vacuum are at maximum. As you move back into the screen, the manifold pressure increases.

It is recommended that you occasionally graph your changes during the calibration process: A picture is worth a thousand words. If you should accidentally make a change that would adversely affect the calibration, it would usually appear as a sharp peak or depression in the contour of the map. See page 59 for how to graph your changes.

Cranking Fuel

When *Cranking Fuel* is chosen, the root menu will disappear and a submenu will take its place on the screen. The Cranking Fuel submenu lists three choices.

Program Cranking Fuel from a File and Save Cranking Fuel in a File—*Program Cranking Fuel from a File* downloads a 16-number table from a file that has been saved on the disk to your ECU. *Save Cranking Fuel in a File* saves these 16 numbers from your ECU to a file on the disk. These values make up the startup cranking fuel table of the ECU. This file specifies the cranking fuel as a function of engine coolant temperature.

Start Up Cranking Fuel is the fuel needed to start the engine. The pulse width is established as a function of the engine coolant during cranking. This selection allows you to monitor and edit the cranking fuel table in the ECU. When you select this function, CalMap reads the contents and displays them in a two-dimensional graph. The vertical axis is the starting engine coolant temperature in degrees Fahrenheit. Also present on the graph are a wedge-shaped pointer and a small box. The pointer is your position controlled by the keyboard of the PC; the box represents actual conditions as seen by the ECU through its sensors.

CalMap continuously reads both the starting coolant temperature and the cranking pulse width value from the ECU and updates the position of the ECU monitoring box. You can change the position of the pointer with the arrow keys on the PC. As you move the pointer in the horizontal (temperature) direction, the display box beneath the graph shows the corresponding vertical (cranking pulse width) value in the table.

Note that CalMap updates this data only while you are cranking the engine. Therefore, if the engine is running, the location of the ECU monitoring box remains where it was during the most recent crank episode. The position of the monitoring box has no meaning when the engine is not running.

Changes are made to this table with the following steps:

• Move the pointer with the arrow keys at the position on the graph that you wish to change. Type the new value. Press the ENTER key.

• After you enter a new value, you may notice a couple of things on the graph that seem odd. First, after you enter a new value in the table, the display box may show a slightly different number from the one you entered. Don't be alarmed. Since the ECU is a digital processor, it has finite resolution, and adjustments are made in small steps. In this table the steps are 0.256 millisecond. After you enter the new value, CalMap converts it to the actual value the ECU will use.

• Second, every time you enter a new number in the table, the graph is redrawn to reflect the change. Notice that when you enter a number that is larger than any value already on the graph, there will be a longer-than-usual pause. This is because CalMap is automatically rescaling the values along the vertical axis so the new point

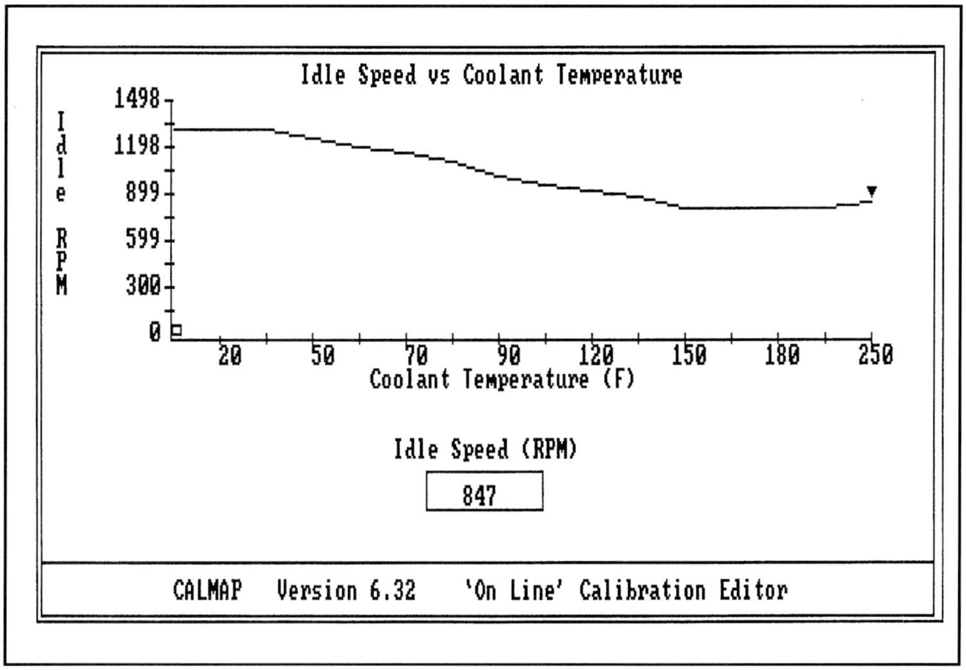

For the idle speed control to function, the throttle plates need to be oriented properly.

of the graph can be seen. It does this by first programming the ECU with the number you have entered, then it rereads and re-plots the table.

Idle Speed

When *Idle Speed* is chosen in the root menu, it will disappear and a submenu will take its place on the screen. The Idle Speed submenu lists six choices.

Program Idle Speed Table from a File & Save Idle Speed Table in a File—If selected, this screen functions as described previously. *Program Idle Speed Tables from a File* downloads four tables, with 16 numbers in each, from a file that has been saved on a disk to your ECU. *Save Idle Speed Tables in a File* saves these four tables from your ECU to a file on the disk. These tables contain values that determine your engine's idle speed, throttle follower, stall saver, and idle air control (IAC) starting position.

Idle Speed—This selection on the submenu allows you to monitor and set the speed of your engine's idle if equipped with an idle air control motor. When *Idle Speed* is selected, CalMap reads the idle speed tables in the ECU and plots a two-dimensional graph of the idle speed table only. The vertical axis of the graph is the engine speed and the horizontal axis is the engine coolant temperature in degrees Fahrenheit. Also present on the graph is a wedge-shaped pointer and a small box. The pointer is your position, which you control with the arrow keys. The box is the ECU's position. To exit the graph and return to the *Idle Speed* submenu, press the Q key.

CalMap continuously reads both the idle speed and the coolant temperature from the ECU and updates the position of the ECU monitoring box. As you move the pointer in the horizontal (temperature) direction, the table value display box, beneath the graph, shows the corresponding rpm. Be aware that, for screen graphic purposes, CalMap limits the vertical (idle speed) position of the ECU monitoring box. So if the engine speed becomes greater than the highest number on the vertical axis, it will not be reflected on the screen.

Changes are made to this table by moving the pointer with the arrow

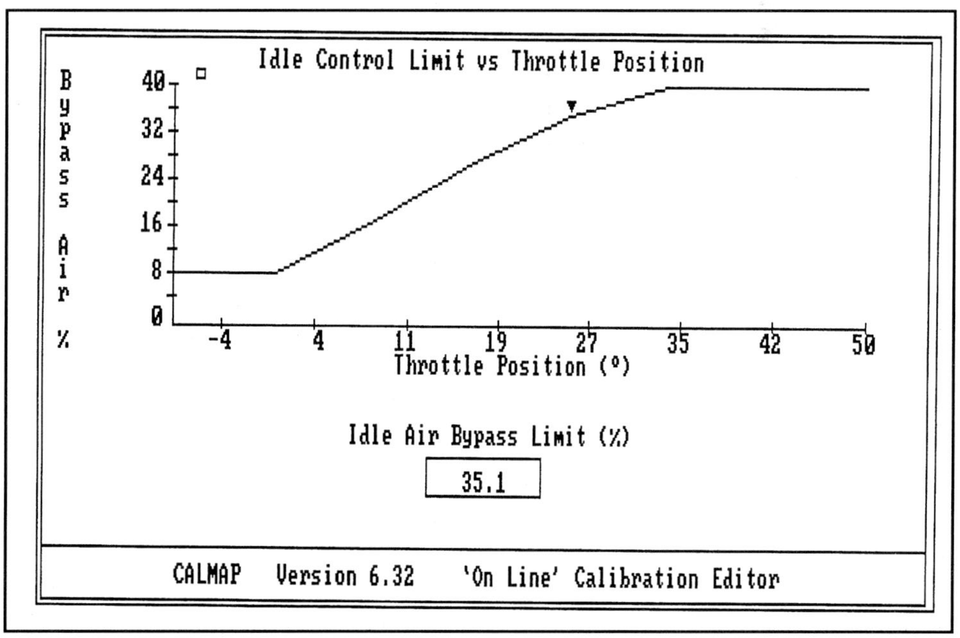

Identified as the throttle follower function, this screen is used to set the amount of air bypassing the throttle plates. Note the location at over 40 percent of the square indicator box. This means that the throttle plates are too far closed and the IAC is fully retracted to keep the engine running.

keys to the position on the graph that you wish to change. Then type the desired value, and press ENTER.

You will notice that after you enter a new value in the table, the display box may show a slightly different number than the one you entered. In this table the adjustments are made in small increments of 7.84 rpm. Second, every time you enter a new value on the graph, it will be redrawn, reflecting the change, and may need to be rescaled.

Throttle Follower—This selection allows you to monitor the throttle position and idle air bypass of the engine. It also allows you to choose the minimum limit of the idle air bypass. If your engine is not equipped with an idle air control motor, this function is not used.

The purpose of the throttle follower is to back the IAC mechanism away from its stop (the fully closed position) in order to allow additional air into the intake manifold at off-idle speeds. This helps to prevent stall under hard deceleration.

When *Throttle Follower* is selected, CalMap reads all of the idle speed tables in the ECU and plots a two-dimensional graph of the throttle follower table only. The vertical axis is the amount of air bypassed around the throttle blades (bypass air percentage), and the horizontal axis is the throttle blade position (degrees of rotation). Throttle position is measured by the throttle position sensor. Also present on the graph are a wedge-shaped pointer and a small box. The pointer indicates your position, which you control with the arrow keys, and the box reflects the ECU status. To exit the graph and return to the *Idle Speed* submenu, press the Q key.

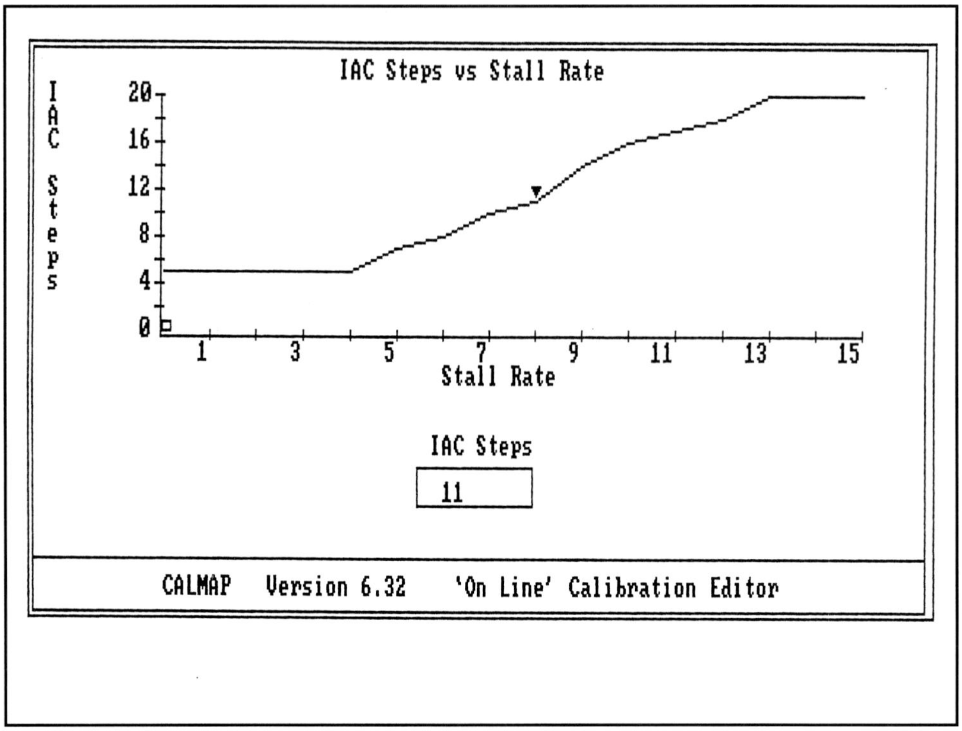

The Stall Saver is used to compensate for a load dump on a slow-running engine.

CalMap continuously reads both the TPS and idle air bypass from the ECU and updates the position of the ECU monitoring box. Since the ECU is a digital processor, it has finite resolution, and calibration changes are made in small increments. In this table, the steps are .392%.

Stall Saver—This feature allows you to fine-tune the driveability when shifting from park to drive. It's also used to maintain a given rpm when accessories are turned on, for example, the air conditioner or engine fan.

The *Stall Saver*, based upon a stall rate (change in rpm and MAP), will open the idle air control motor a specified number of steps to prevent the engine from stalling. These steps occur immediately. Once they are triggered, the IAC will remain in the trigger position momentarily. It then resumes its normal operation.

When *Stall Saver* is selected, CalMap reads all of the idle speed tables in the ECU and plots a two-dimensional graph of the *Stall Saver* table. The vertical axis is the number of steps the IAC moves and the horizontal axis is the stall rate. Also present on the graph are a wedge-shaped pointer and a small box. The pointer indicates your position, which you control with the arrow keys. To exit the graph and return to the *Idle Speed* submenu, press the Q key.

CalMap continuously reads both the IAC steps and the stall rate from the ECU and updates the position of the ECU monitoring box. You can change the position of the pointer with the arrow keys on the PC. As you move the pointer in the horizontal (stall rate) direction, the table value display box beneath the graph shows the corresponding vertical (IAC steps) value in the table for your pointer's stall rate location. CalMap responds to changes in the Stall Rate calibration in steps of 0.392%.

IAC Starting Position—This

function allows you to program the initial starting position of the IAC for a given coolant temperature. This mode is activated only when starting the engine. Once the engine is running, the IAC motor defaults back to maintain the rpm of the Idle Speed curve.

When the *IAC Starting Position* is selected, CalMap reads all of the idle speed tables in the ECU and plots a two-dimensional graph of the *IAC Starting Position* table. The vertical axis is the percent of air bypassed by the IAC, and the horizontal axis is the starting coolant temperature. Also present on the graph are a wedge-shaped pointer and a small box. The pointer is your position, which you control with the arrow keys, and the box indicates the ECU status.

CalMap continuously reads the Bypass Air Percentage and the startup engine coolant temperature from the ECU, and updates the position of the ECU monitoring box on the screen. You can change the position of your pointer with the arrow keys on the PC. It is important to note that the ECU updates the startup temperature values only during cranking. Consequently, CalMap does not show any monitoring box movement in the horizontal direction after the engine starts. The ECU responds to changes in the calibration made with the CalMap software in steps of 0.787%.

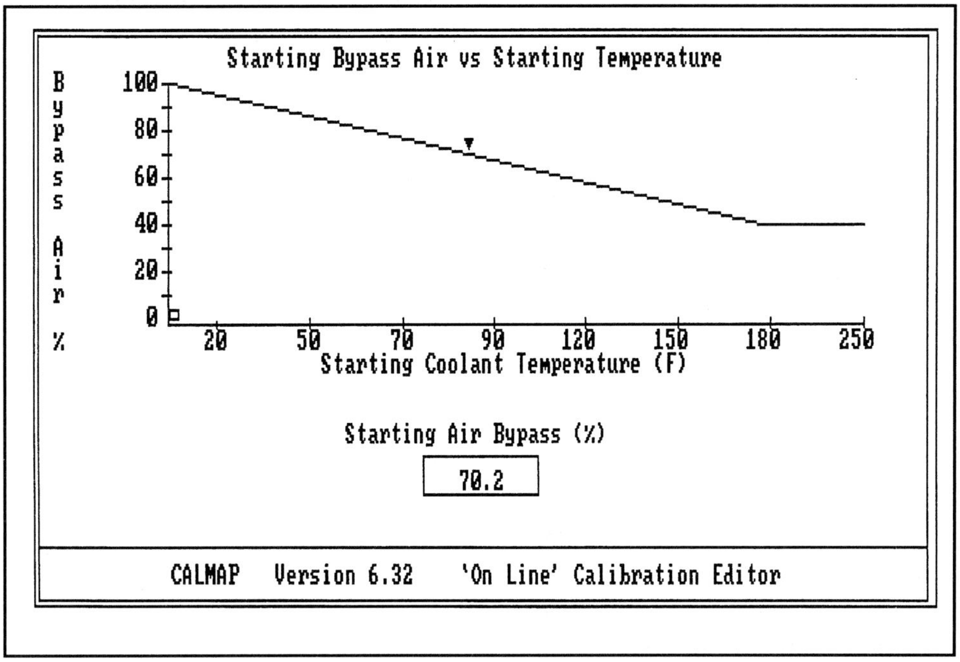

This screen controls the IAC position on startup.

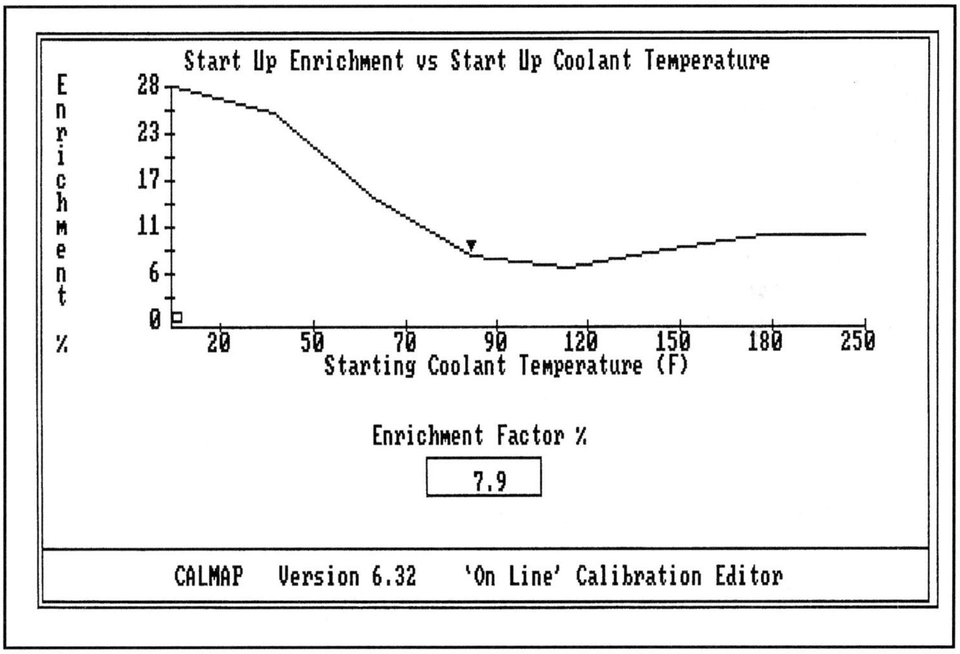

This is the additional fuel added over the value in the base fuel map during Afterstart Enrichment.

Afterstart Enrichment

The Afterstart Enrichment factor is intended to help overcome the initial startup frictions and viscosities present in a cold engine. It is also designed to compensate for hot soak fuel vaporization in the fuel system of hot engines. The *Afterstart Enrichment Tables* submenu lists five choices.

Program the Afterstart Tables from a File downloads three tables with 16 numbers in each to your ECU from a file that has been saved on disk. Selecting *Save the Afterstart Tables in a File* saves these three tables from your ECU in a file on the disk. The tables contain the values that determine the Afterstart fuel enrichment characteristics.

Initial Afterstart Enrichment— This submenu selection allows you to monitor and change the amount of Afterstart Enrichment added to the base fuel quantity delivered to the

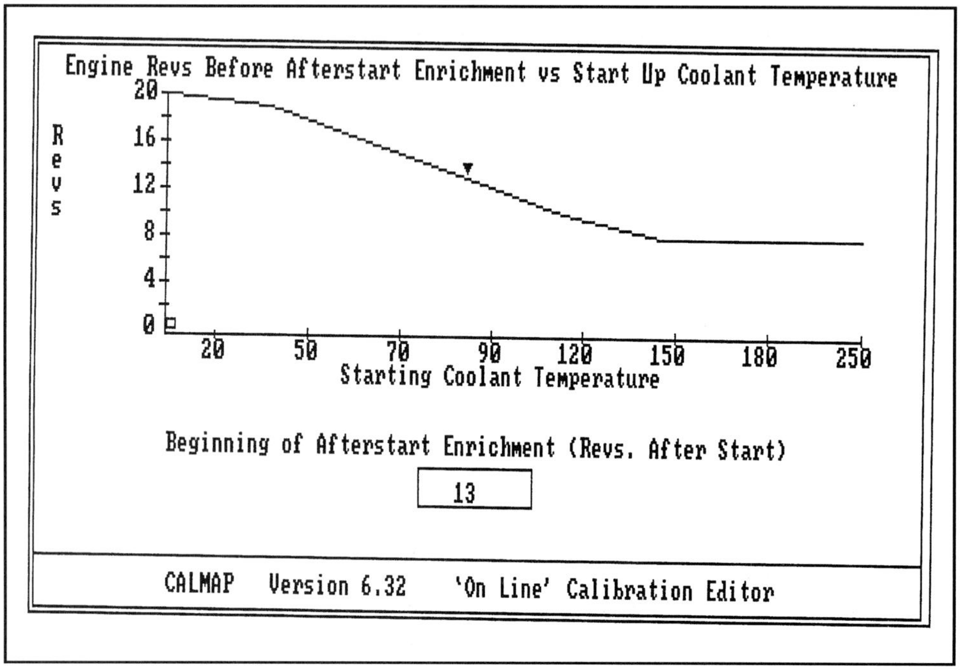

This table allows you to decide when the Afterstart Enrichment is initiated.

engine. When this table is selected, CalMap reads the Afterstart Enrichment tables in the ECU and plots a two-dimensional graph of the *Initial Afterstart Enrichment.* The vertical axis indicates the percentage of enrichment, while the horizontal axis represents the engine coolant temperature at startup. Also present on the graph are a wedge-shaped pointer and a small box. The pointer is your position, which is controlled with the arrow keys on the PC, and the box indicates the ECU status.

CalMap continuously reads the Afterstart Enrichment and the startup engine coolant temperature from the ECU and updates the position of the ECU monitoring box on the screen. You can change the position of the pointer with the arrow keys. As you move the pointer in the horizontal direction (startup temperature), the table value display box beneath the graph shows the corresponding vertical value (Initial Afterstart Enrichment in percentage) in the table for your pointer location. The ECU updates the startup temperature value only during engine crank. Changes to the calibration of this table with the CalMap software are in steps of 0.787%.

As the submenu name implies, the *Afterstart Enrichment Table* contains enrichment factors that are present only for a short time after startup. This enrichment factor is added after a certain number of engine revolutions as specified in the *Beginning of Afterstart Enrichment* table.

The Afterstart Enrichment decays to zero at a rate specified in the *Afterstart Enrichment Decay Rate* table. When you view this on the initial *Afterstart Enrichment* graph (as the engine is running), you will see that at startup the monitoring box rests on the horizontal axis (0%) at the current coolant temperature. Then, when the engine has turned the number of revolutions specified in the *Beginning of Afterstart Enrichment* table, the box will jump to the set point on the graph, signaling that the ECU has put this factor of enrichment in.

As the engine continues to run, the monitoring box will fall in 0.8% steps at the rate specified in the *Afterstart Enrichment Decay Rate* table until it reaches zero percent enrichment.

Beginning of Afterstart Enrichment—When this is selected, CalMap reads all of the Afterstart Enrichment tables in the ECU and plots a two-dimensional graph of the delay revolutions until Afterstart Enrichment is invoked. The vertical axis is the engine coolant temperature at startup. Also present on the graph are a wedge-shaped pointer and a small box. They function similar to the other tables by way of the arrow keys.

CalMap continuously reads the number of engine revolutions from startup and the engine coolant temperature at startup from the ECU, and updates the position of the ECU monitoring box. You can change the position of the pointer with the arrow keys on the PC. As you move the pointer in the horizontal direction (startup temperature), the table value display box beneath the graph shows the corresponding vertical value (number of engine revolutions after engine startup) in the table for your pointer's startup engine coolant temperature location. The ECU updates this information only when the engine is cranking.

When making calibration changes using CalMap, note that the software is coded for each step as two revolutions of the crankshaft. Immediately after the engine has been started, the monitoring box will begin to fall from its starting position on the Initial Afterstart Enrichment graph. The faster the engine speed, the faster the box will fall. When it reaches the horizontal axis, zero, the ECU will no longer add the enrichment. This table is a factor of the values in the Base Fuel Map. The box in the beginning of the Afterstart Enrichment graph will begin to cycle between the specified rpm value and zero until the enrichment has been timed out. This reflects the cycling of the Afterstart Enrichment revolution counter in the ECU. Every time the box reaches zero, the enrichment is decreased by 0.8% until it reaches zero enrichment.

Afterstart Enrichment Decay Rate—This table allows you to monitor and change the number of engine revolutions that occur between each Afterstart Enrichment factor decay step. This function has the effect of controlling the smoothness of the transition from Afterstart to normal engine operation.

When *Afterstart Enrichment Decay Rate* is selected, CalMap reads all of the Afterstart Enrichment tables in the ECU and plots a two-dimensional graph of the Afterstart Enrichment Decay Rate only. The vertical axis is the number of engine revolutions between each Afterstart Enrichment Decay step. The horizontal axis is the engine coolant temperature at startup.

Also present on the graph are a wedge-shaped pointer and a small box. The pointer shows your position, which is controlled with the arrow keys, and the box is the ECU's. As with the other tables, CalMap continuously reads the number of engine revolutions from startup and the startup engine coolant temperature from the ECU, and then updates the position of the ECU monitoring box.

As you move the pointer in the horizontal direction (startup temperature), the table value display box beneath the graph shows the corresponding vertical value (number of revolutions per decay step) in the table for your pointer's startup temperature location. The ECU updates this information only when the engine is cranking.

After the engine has been started, the monitoring box will begin to fall from its starting position on the *Initial Afterstart Enrichment* screen. The faster the engine speed, the faster the box will fall. When it reaches the horizontal axis (zero revolutions) for the first time, the ECU will no longer add the enrichment factor. The monitoring box of the *Afterstart Enrichment Decay Rate* screen will cycle between the specified rpm value and zero. Every time the monitoring box reaches zero, the *Initial Afterstart Enrichment* is reduced by 0.8% until it reaches zero enrichment. At that time the box continues to cycle but there is no effect on the fuel flow.

The *Afterstart* tables are some of the most confusing of the entire DFI system and can be summarized in the following way:

The *Afterstart Enrichment* is a transitional fuel enrichment. It is divided into three parts in order to fine-tune the transition between *Cranking Fuel* and *Warm-Up Enrichment*. The first part tells the ECU how much fuel to add for a given coolant temperature (Initial Afterstart Enrichment). The second part tells the ECU the rate at which to add the enrichment (Beginning of Afterstart Enrichment). The third part tells the ECU the rate to start to remove the enrichment (Afterstart Decay Rate). Although this is a simplified explanation, the use of these tables will be covered more thoroughly in Chapter 8

Warm-Up Enrichment—Program Warm-Up Table from a File downloads the table, made up of 16 numbers, that has been saved on disk to your ECU. *Selecting Save Warm-Up Table in a File* saves this table in your ECU in a file on the disk. The table contains the values that determine the engine's warm-up fuel-enrichment characteristics.

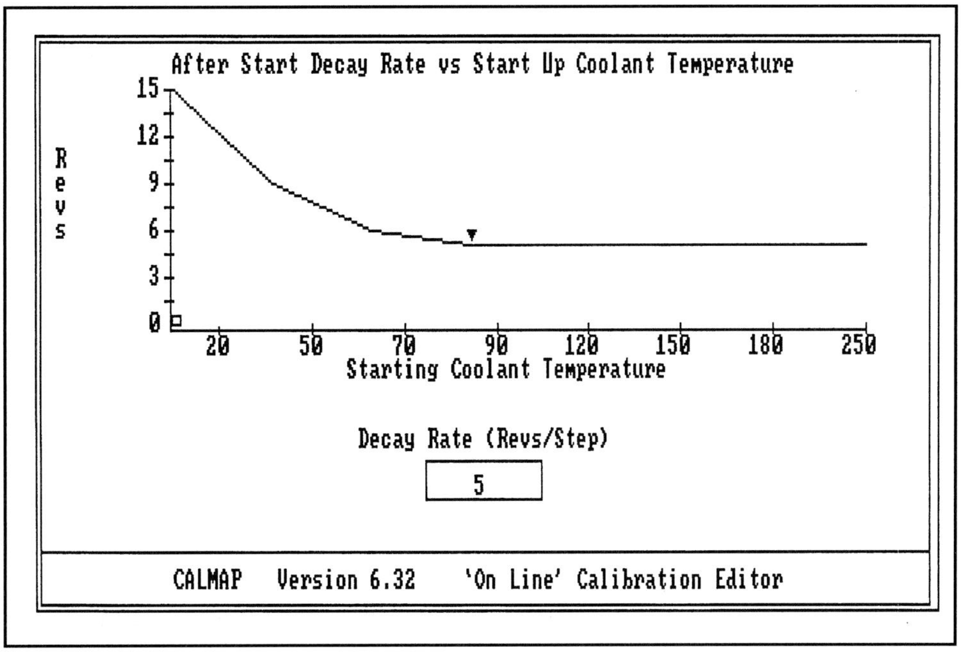

In contrast, the decay rate controls the removal of Afterstart fuel.

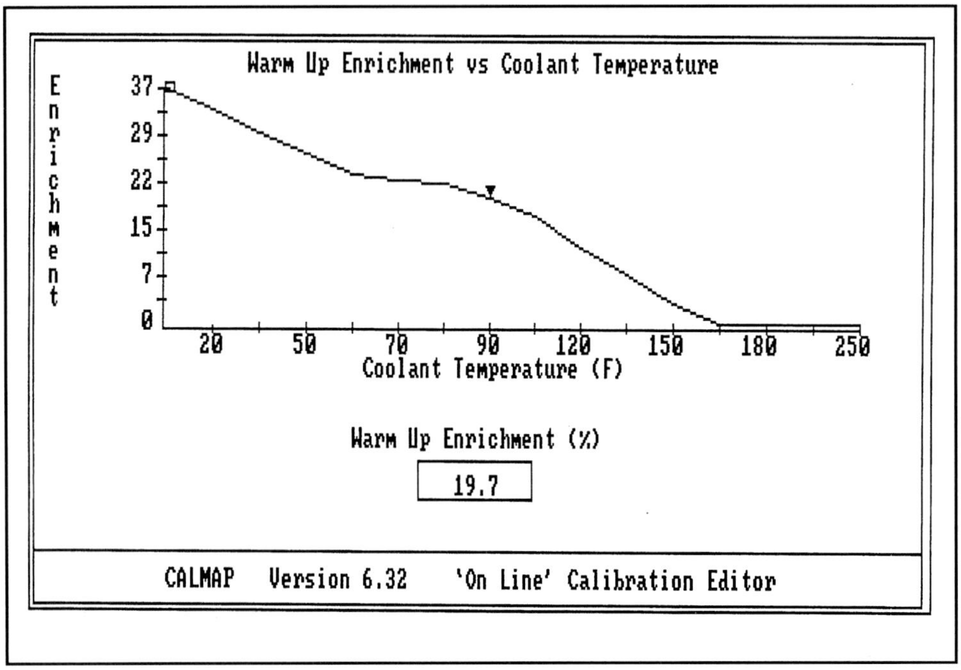

The Warm-Up enrichment can be compared to the opening rate of a thermostatic choke spring on a carburetor.

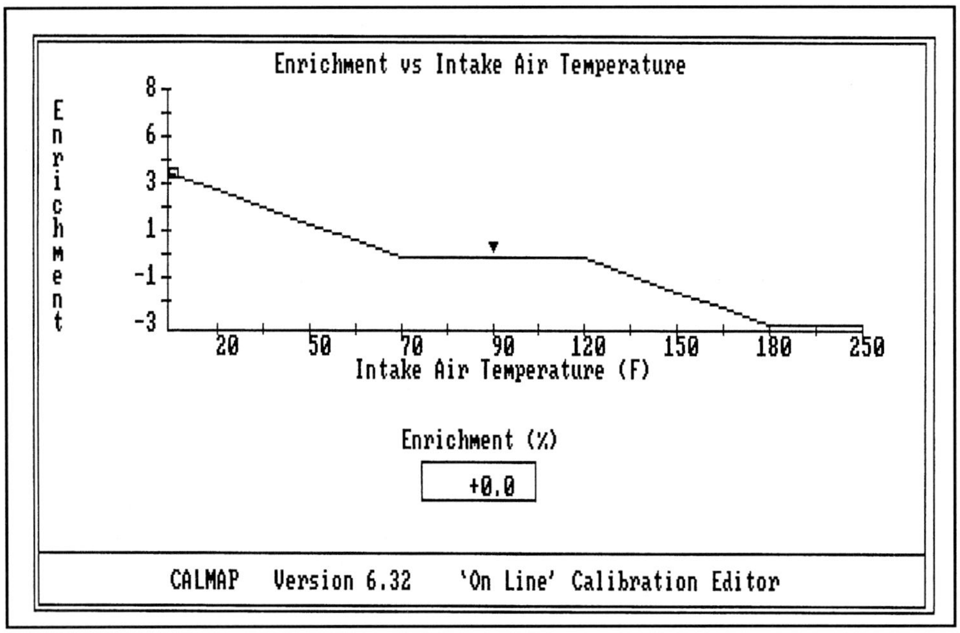

The density of the incoming charge air from temperature change can be compensated for with a few key strokes.

The *Warm-Up Enrichment* selection allows you to monitor and change the amount of enrichment of the base fuel pulse during the engine's cold operation. When this table is chosen, the CalMap reads the *Warm-Up Enrichmen*t table and plots a two-dimensional graph. The vertical axis is the percent of enrichment, while the horizontal axis is the engine coolant temperature in degrees Fahrenheit. In this table the steps are 0.787% when the CalMap converts any inputted tuning changes.

As the name implies, this table is used as a choke function to richen the air/fuel ratio as a function of the value in the Base Fuel Map.

Air Temperature Correction

Program Air Temperature Correction from a Table downloads 16 numbers that have been saved on disk to your ECU. Selecting *Save Air Temperature Correction Table in a File* saves this table from your ECU to a file on the disk. The *Air Temperature Correction* table contains values that determine the amount of fuel to add or subtract from the base fuel delivery as a function of the engine's intake air temperature.

The vertical axis of the graph is the percent of air temperature correction, while the horizontal axis is the engine intake air temperature in degrees Fahrenheit. Inputted changes to the software will be adjusted to this table at a rate of 0.787% for each step. CalMap will replot and rescale the graph as required.

Acceleration Enrichment

The *Acceleration Enrichment* tables control transient fueling. This is the ACCEL software's replacement for a carburetor's acceleration pump circuit. The *Acceleration Fueling* curves determine the additional amount of fuel to be injected, based on the rate of throttle manifold pressure change (TPS and MAP).

Program Acceleration Tables from a File downloads two tables, with 16 numbers in each, to the ECU from a file that has been saved on disk. *Save Acceleration Tables in a File* saves these two tables from your ECU to a file on the disk. The tables contain the values that determine your engine's acceleration enrichment characteristics.

Selecting TPS Acceleration Fuel allows you to monitor and change the amount of fuel enrichment added asynchronously (independent of the base fuel pulses) based on how fast the accelerator pedal is depressed. When *TPS Acceleration Fuel* is selected, CalMap reads the acceleration enrichment tables in the ECU and plots a two-dimensional graph of the acceleration enrichment. The vertical axis is the pulse width in milliseconds, which is added during acceleration, and the horizontal axis is the rate of change in throttle position (TPS). Although the units that are displayed on the horizontal axis (rate of TPS change) are from 0 to 15, the real meaning is: 0 = steady state, the throttle is not moving at all; 15 = the quickest rate of throttle position change. Everything between 0 and 15 is proportional to the two endpoints.

Selecting *MAP Acceleration Fuel* allows you to monitor and change the amount of fuel enrichment added to the base fuel pulses as a function of acceleration sensed by a change in manifold pressure. This function was added to improve driveability. It can be compared to the power valve circuit on a carburetor. A small change in throttle position may result in a large change in airflow through the throttle body, especially with larger throttle bodies or when the throttle body linkage is not progressive or staged. Without sensing the change in manifold pressure, the change in throttle position may not be sufficient to cause the ECU to richen the mixture adequately, resulting in a stumble or bog. The change in manifold pressure will be detected by the MAP sensor, though.

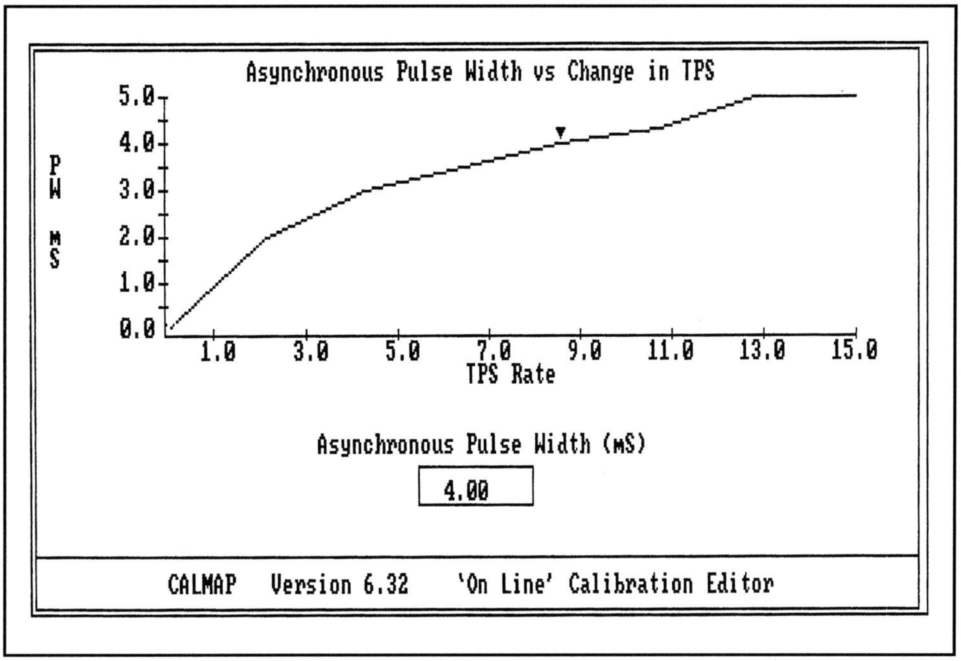

This table can be likened to tuning an accelerator pump circuit in a carburetor.

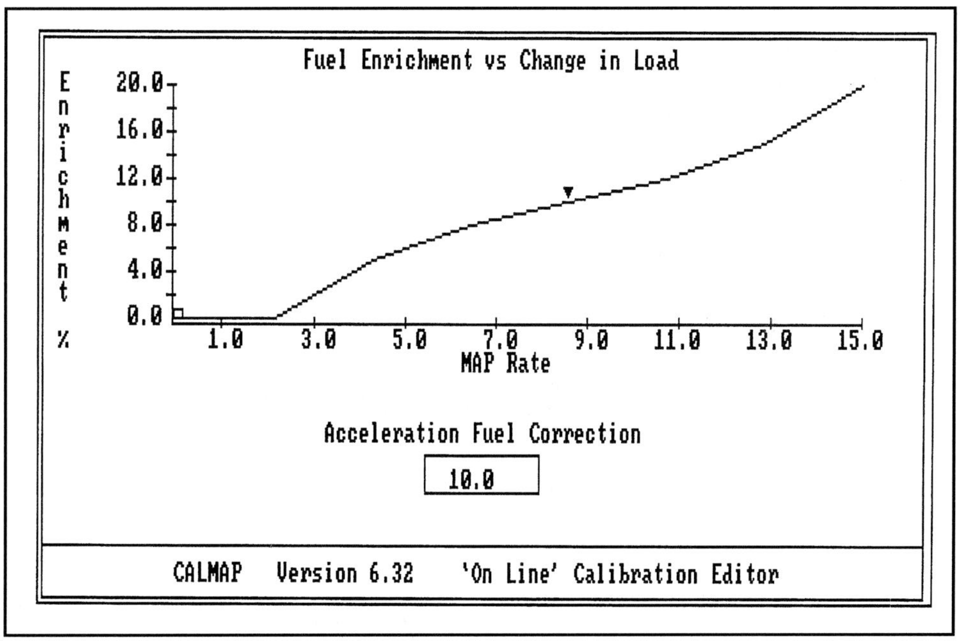

Based on changes in engine vacuum, the MAP sensor functions in a similar way to a power valve.

When this table is selected, CalMap reads the acceleration enrichment tables in the ECU and plots a two-dimensional graph of the acceleration enrichment function of sensing the decrease in manifold pressure. The vertical axis is the pulse width in milliseconds which is added to the base pulse width during acceleration, and the horizontal axis represents the rate of change in manifold pressure (vacuum). Although the units displayed on the horizontal axis (rate of manifold pressure change) are from 0 to 15, the real meaning is: 0 = steady-state, manifold pressure is not changing; 15 = the fastest possible manifold pressure change. Everything

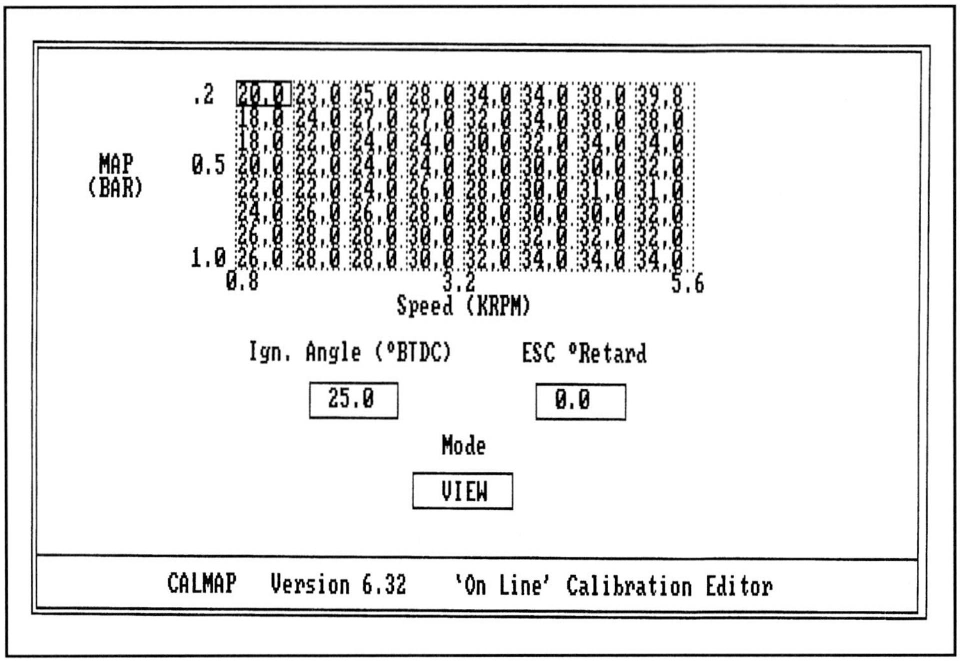

The timing matrix is scaled at a ratio of 1.5 to the fuel map.

between 0 and 15 is proportional to the two endpoints.

Since the ECU is a digital processor, it has finite resolution, and adjustments are made in small increments. In this table, the steps are 0.032 milliseconds.

Ignition

The *Ignition* submenu lists four choices. The first two, if selected, function as described previously. When you enter *Program Ignition Map from a File,* a matrix of 64 numbers is downloaded to the ECU from a file that has been saved on a disk. When you enter *Save Ignition Map in a File*, this matrix of numbers is saved from the ECU to a file on the computer disk. The matrix of numbers makes up the *Ignition Map* for the engine controller and determines the ignition timing at each possible speed/load operating point on the engine.

The *Ignition Map* selection lets you monitor and edit the basic timing characteristics of the engine whether it is running or not. Once this selection is made, the screen will clear and then begin filling with a matrix of numbers. This matrix is the ECU's timing table and will be placed in the upper portion of the screen. There will be eight rows and eight columns of numbers. The rows correspond to the manifold absolute pressure, the columns to the engine speed.

The middle of the screen will contain the interpolated ignition angle. This is the total angle interpolated from the Ignition Map. It should be synchronized with the timing mark on the engine damper.

The lower portion of the screen shows the total ignition angle retard. This reading should be zero when the optional knock sensor kit is not used. When the knock sensor is used, depending on the amount of detonation, the actual spark retard will be shown in this box.

The location of the flashing cursor box in the ignition indicates the cell in which the ECU is currently operating. The flashing cursor box correlates to the load, or manifold absolute pressure, and engine speed based on the sensor inputs. If the engine is not running, the flashing box will be in the lower right corner. During normal operation the box will move to the right. As manifold pressure increases, it will move downward.

Because the Ignition Map is an 8x8 matrix and the Base Fuel Map is a 15x16 matrix, you will notice a scaling difference of 1.5 between the two maps' cell widths. Therefore, when the ECU is configured for 400-rpm increments, in the Configure ECU mode, the Base Fuel Map cell widths will reflect 400-rpm increments while the Ignition Map cell widths will reflect 600-rpm increments (600 = 400 x 1.5). The Ignition and Base Fuel Maps are very similar to one another. In the Base Fuel Map, an injector pulse width is being generated for a given rpm and manifold pressure. In the Ignition Map, a spark angle is being generated for a given rpm and manifold pressure.

Graph Ignition Map—When this is selected from the menu, a bar graph on the screen showing the progress of the upload will indicate when the process is complete. After all of the data is received by the PC, a three-dimensional plot of the Ignition Map will be generated and displayed. The point on the plot where the vertical axis meets the horizontal axis is where both the ignition angle and rpm are 0. As you move up the vertical axis, the ignition angle increases. As you move to the right on the horizontal axis, the engine speed increases. The point where the horizontal axis meets the axis going back into the screen is where the engine speed and the manifold vacuum are maximum. As you move back into the screen, the manifold pressure is increasing.

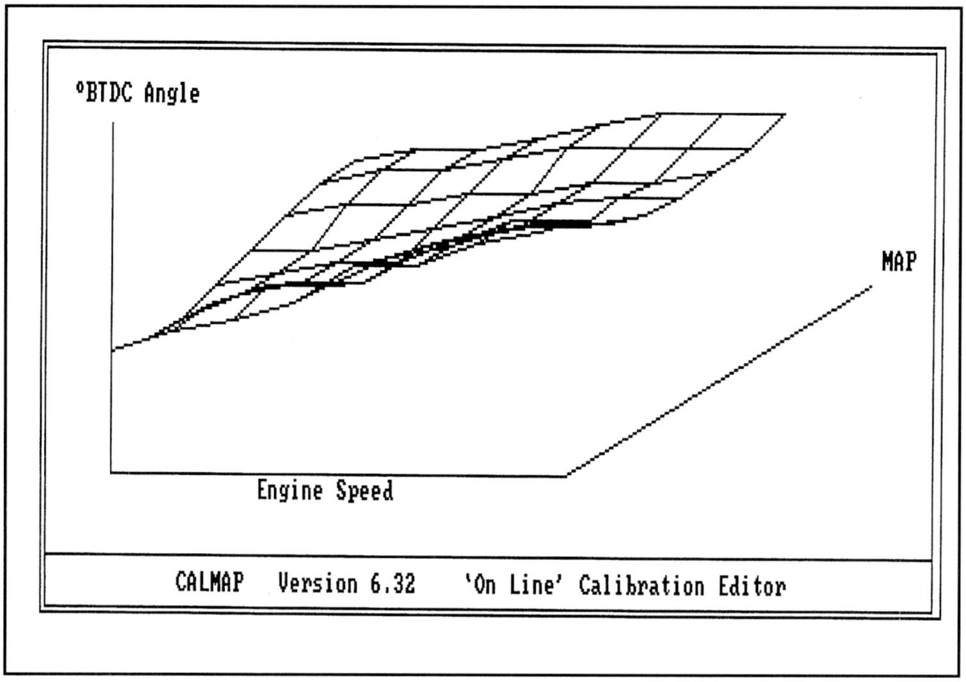

Graphing the timing map offers the same advantage as looking at the fuel matrix in this form.

It is recommended that you occasionally graph your changes during the calibration process. If you should accidentally make a change that would adversely affect the calibration, it would usually appear as a sharp peak or depression in the contour of the Ignition Map.

Global Submenu

This mode contains additional controls that allow you to customize the ECU for your specific application. In addition to letting you program or save all of the ECU's tables simultaneously, this submenu lets you preset the number of cylinders, Base Fuel Map cell size, manifold pressure range, torque converter lockup conditions, rev limit, and coolant fan operation. It also allows you to specify whether open- or closed-loop operation is desired, and the conditions for operating in closed loop. The Global submenu also contain nitrous oxide control programming capabilities, which will be full discussed in Chapter 8.

Program All ECU Tables from a File

When this is selected, you may program new table values into the ECU from a file that has been previously saved by CalMap. First, a list of the files in the directory will appear. These files are your only choices for programming the ECU at this time. Once you have created more files, the number of choices from the directory will grow. After the files have been listed, you will be asked for the name of the file you wish to program into the ECU. If you decide not to program a new file into the ECU, just press the ENTER key without typing in the name of a file and you will be returned to the submenu from which you started. To program a file into your ECU, type in the name of the file you wish to use and then press ENTER. CalMap will begin programming the ECU from the file and then return you to the submenu from which you started. If CalMap has a problem finding the file you have requested, it will honk at you and display an error message. Press the Q key and you will be returned to the submenu for another try.

Save All ECU Tables in a File

When this is selected, a bar graph will appear on the screen to notify you of the progress of the reading process. Once the table data is read from the ECU, the screen will list all of the previously saved tables. Above the file list, a question will appear asking you to name the new file. Type in the name you wish to assign to this new file. Unfortunately, there are limitations to the naming of files. File names can contain only letters and numbers (alphanumeric characters), and there can be no more than eight characters in the name. However, file names can also have up to three character extensions. Some examples of files names are: 305280, 350.IDL, 383Turbo.MAP, etc., where the extension is the set of characters following the dot (period). Once you have typed the new file name, press the ENTER key to save it.

After the file has been saved, you will be automatically returned to the corresponding submenu. If, after the table has been read from the ECU, and the file list has been displayed, you decide that you do not want to save the table after all, do not enter a file name. Just press the ENTER key and you will be returned to the submenu without having saved the data in a file.

Record Engine Operating Data

When this menu item is selected, CalMap reads the appropriate

```
*  Coolant Temp.                    Throttle Position
   Coolant Correction            *  Base Pulse Width
   Air Temp.                     *  Engine Speed
   Air Correction                   ESC Spark Retard
*  Ignition Voltage                 Additive Pulse Width
   Ignition Correction              Multiplicative Correction
   Manifold Pressure                IAC Position
*  O2 Voltage                       Gross Pulse Width
   O2 Correction                    Spark Angle

   Use the arrow keys to highlight a selection then press the
   [ENTER] key to mark/unmark the selection.  An asterisk will
   appear in front of marked items.  Up to six items may be
   selected. Press the [L] key when finshed with selections.

        CALMAP   Version 6.32    'On Line' Calibration Editor
```

Up to six different parameters can be logged for review as a tuning aid. This function works very well for drag strip tuning.

memory file from the disk in your computer. This file contains the locations in your ECU's memory for all the data locations: battery voltage, engine speed, air temperature, etc. These locations contain the actual data that your ECU uses during operation. When you log data, you see exactly what your ECU sees.

The first thing the *Data Log* function does is prompt you with a screen full of various ECU inputs and outputs. Use the arrow keys to move from selection to selection. Press the ENTER key to tag a selection for logging. Press it again to untag it. Up to six data points can be selected for logging. Once you have selected and tagged the data points, press the L key to log the data. Press the Q key to exit the data-logging mode. This will then return you to the Global submenu.

After the data points to be logged are selected, the screen will clear and a log mode selection will be displayed. Your choices in this screen are *Log to Disk* or *Snapshot*. The *Log to Disk* selection is designed for slower sample rates. Data is collected based on a sample rate and saved on the disk as it is collected. The *Snapshot* mode is designed for much quicker sampling. Data is collected as fast as CalMap can get it. Use the right/left arrow keys to pick a mode, and then press ENTER to select it.

If the *Log to Disk* mode is selected, you will be prompted with a screen that allows you to pick the rate of sampling (how often CalMap will collect the data from the ECU). Use the arrow keys to move the selection box to the sample rate you desire, and then press ENTER. Once the sample rate has been selected, the screen will clear and then display a list of the previously saved data files you have created. Above the file list, you will be asked for the name of the file for the collected data to be stored. Enter a name and press the ENTER key to begin logging data. After the file name is entered, the screen will reconfigure to display the data that is being collected. This allows you to monitor engine functions in real time, as well as save the data for later inspection. When you wish to terminate the data-logging mode, press the Q key.

Some facts to remember: If the PC is rebooted or the power is lost during logging, the data will not be retained in the file. If the disk that the data is being saved on becomes full during the logging process, the data will not be retained in the file. Therefore, if there are a lot of old data files in the directory, and there is a chance that the disk may become too full, it may be best to first go into DOS and delete some of the old data files. This may save you a lot of headaches later.

If the *Snapshot* mode is selected, the screen will clear and then display a list of the previously saved data files you have created. Above the file list you will be asked for a file name to store the data. Type in a file name and press ENTER.

After typing in a file name, you will be prompted with a screen that allows you to pick the trigger for the initiation of the data collection. Use the arrow keys to move the selection box to the trigger you desire, and then press the ENTER key. The trigger is the ECU data that is monitored continuously. It can be the engine speed, manifold pressure, throttle position, or manual trigger. Once the trigger is selected, you will be asked for the point at which to start gathering data. For example, you may want to start collecting data at 1,000 rpm. If so, type in 1,000. If you select a manual trigger, the data-collecting mode is cued by the ENTER key. Once ENTER is depressed, the data is collected for 15 seconds. The screen will go blank until all the data is accumulated. Once finished, the screen will indicate it is saving data in file mode.

Because there are so many ways to analyze and view the data you have recorded, CalMap saves the data in a format that is easily readable and portable into other programs, standard ASCII. There are no provisions in CalMap to read the data in these files from within CalMap. If you wish to view the data, you must do it in DOS, a screen editor, or read it into some other program. To view the data from DOS, you must exit the CalMap program and return to DOS. To view the data type: TYPEDATA/ filenameENTER (all one word with no spaces). The data will begin scrolling up the screen. To stop the display, hold the CTRL key down and press the S key. To continue, hold the CTRL key down and press the S key again. To print the data, assuming your printer is connected properly, type: PRINTDATA/filenameENTER. Screen editors like BSE or WordStar, spreadsheet such as Lotus 123, and custom programs of your own can also be used to read and analyze the data you have recorded.

CalMap Version 6.32

This is the latest and final version of CalMap offered to the public. There were a few Version 6.33 disks in the hands of some dealers, but that was more of an upgrade to function with newer PCs than an alteration to the engine-management software. Version 6.32 will be required to run on any PC that is faster than a 280, which means anything built within the past seven or eight years. Versions 6.0, 6.1, and 6.2 were too slow to operate on faster computers and caused many lockup conditions. If you are using an old DOS-based computer, it will function best with any software version other than 6.32.

Version 6.32 does have additional features to help tune the engine that were not offered in earlier software. By striking the backslash key while in the fuel map, the matrix values chosen by the ECU will become highlighted. No longer will you have to try to follow the cursor while the engine sweeps through different locations. By depressing the backslash key again, the function is canceled. In the Global menu there is a new feature under the Utilities submenu; it has the ECU create a base fuel map from information about injector sizing, horsepower, torque, etc. This function works very well at blending the fuel values before and after the torque peak and is so easy to select that it should be used on every new calibration. The author's experience has shown that this program does drift to the rich side of an acceptable mixture, but it's meant to be only a starting point for your calibration, not a final tune-up.

Edit ECU

This function was incorporated to aid in offline calibration table changes. With it you have the ability to alter all the submenus of the *Edit ECU* mode except for the Global submenu. It is especially beneficial when making major changes to the Base Fuel Map. In the *Edit File* mode, the Base Fuel Map can be modified by multiplying, adding, or subtracting coefficients to increase or decrease the fuel flow. This function also allows the editing to be done offline (not connected to the ACU). The file that is to be edited must, however, be saved on the disk in the Global directory. A good example of when this function would be used is: Suppose you are going to raise the fuel rail pressure but you do not want the engine to run richer. If you knew the percent increase in fuel flow based upon your new fuel rail pressure, you could rescale the complete fuel map with a few key strokes instead of changing each of the individual cells, which is very cumbersome. Using the arrow keys will allow you to move around and edit individual cells if you wish.

Keep in mind that a multiplicative shift will increase the whole map up or down by a certain percentage, whereas the addition and subtraction tend to have a greater effect on the smaller numbers in the fuel map. For example, if you were to add five counts to the entire map:

If the initial value in the map at idle was the number 30, adding five counts increases it to 35, which is about a 15-percent increase. However, adding five counts to the number 100 increases the fuel delivery time by only five percent. It is quite obvious that the addition and subtraction will have a greater effect at small pulse width values.

To use this function, first verify the file that you wish to edit is stored in the Global directory. To do this go to the Global menu in CalMap. Then go to *Program All ECU Tables from a File* and make sure the file you wish to edit appears on the screen. If not, hit the ENTER key and go to *Save All ECU Tables in a File* and save the file you wish to edit. Once you have verified the file you wish to edit is saved on the disk, go to CalMap's first screen. Now move the arrow key to EDIT FILE and press ENTER. The next screen will ask you the *ECU Configuration Code Number*. Type in 015013 and then press ENTER. The next screen to appear will ask the name of the Global file you wish to edit. The file names saved should be displayed on the screen to help you remember what files are saved. Once you enter the file name, press ENTER. The next screen to appear will be similar to the root menu of the *Edit ECU* mode, but the Global submenu will be missing. You can not change the submenus of the Global section in the *Edit File* mode. All of the submenus of the root menu (*Cranking Fuel, Idle Speed, Afterstart Enrichment, Warm-Up Enrichment, Air Temperature Correction, Acceleration Enrichment*, and *Ignition*) can be modified as described in the *Edit ECU* mode section. However, the Base Fuel Map menu can be modified even more in the *Edit File* mode than in the *Edit ECU* mode.

Modifying the Base Fuel Map—As was mentioned previously, the Base Fuel Map can be modified by multiplying, adding, or subtracting coefficients to increase or decrease the fuel delivery by the factors used to modify the Base Fuel Map. To use this feature, move the box with the arrow keys to surround the Base Fuel Map submenu of the root menu. Once it's surrounded, press the ENTER key. The Base Fuel Map will appear. If you want to multiply the complete matrix by some number, type M. The screen will now ask you the percent change you wish to shift. One hundred percent is unity, so if you wish to increase the map by 10 percent, type 110%. Likewise, if you wish to decrease the map by 10 percent, type 90%. If you wish to add or subtract counts, type A for add, S for subtract. The screen will ask you to enter the number of counts to modify the fuel map by and will completely rescale itself in a few seconds.

Graphing the Base Fuel Map—If you wish to graph the Base Fuel Map, type G. The matrix of numbers will disappear and a three-dimensional polygon mesh will appear. This graph functions in this mode very quickly because CalMap does not have to communicate with the ECU to read the data.

Modifying Other Submenus—If you wish to modify any other submenus of the root menu, you can do so just like you did in the Edit ECU mode. Once you have completed your work, press the Q key until you return to the root menu of the *Edit File* mode. Once you return to the root menu, press Q. CalMap will now ask for the name of the file in which to save the revised calibration. At this point, type in a file name you want to save it under and press ENTER. An example would be 355REEV1.ALL. If you decide to again modify these tables, you might change the new revision to 355REV2.ALL, etc.

After typing the file name, press ENTER. The CalMap screen will then indicate that it is saving the Global file you just modified to the CalMap disk. The screen will then default back to the original ACCEL logo screen with the *Edit ECU/EDIT File* modes. It is important to remember that the changes made in the *Edit File* mode were saved to a Global file on the disk. In order to put these changes into the ECU, you have to get back into the *Edit ECU* mode and download the file from the disk to the ECU.

7

Working with CalMap Software

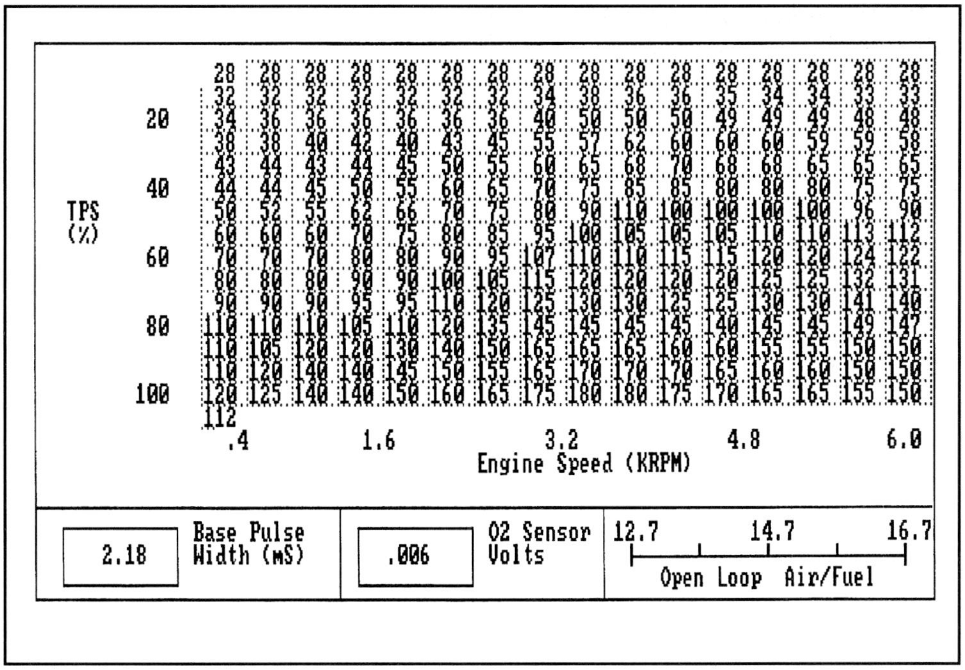

ACCEL's CalMap software, in conjunction with the company's ECU, was the first truly affordable user-friendly programmable fuel-injection system. CalMap is partially responsible for the acceptance and popularity of today's new breed of performance cars.

The CalMap software is a proprietary program that will interface only with ACCEL ECUs and was the first aftermarket program that allowed on-the-fly tuning. When ordering ACCEL part number 74990-L or 74990-S, you receive a 3.5-inch disk, an instruction manual, and an interface connector with a write protector. The part number designation identifies the interface cable length of either five or 25 feet. The longer of the two cables is usually purchased for dyno cell use, with the standard five-foot length being sufficient for almost all applications. The end of the interface cable that attaches to the ECU main wiring harness uses a Packard four-pin, connector while the IBM-compatible laptop connection incorporates an RS232 fixture. With this software and serial cable, you can change any or all of the available 256 points in the fuel table, along with the ignition timing, idle speed, enrichment circuits, and configuration. The software package allows access to an unlimited number of ACCEL ECUs and needs to be purchased just once for use on a multitude of applications. The original software purchase does not entitle the owner to any free software upgrades though. In this instance, only a new disk would be required.

Due to the unique nature of the ACCEL ECU, it does not need to be removed from the vehicle to be reprogrammed or to calibrate any of the tables. They can all be accessed with either the engine running or the ignition on and the engine off. There are no PROMS to reburn and no additional expense for software needed after the original investment.

Working with CalMap

After booting up your PC, at the DOS prompt, type in the word CalMap (the software is not sensitive to capitalization) and wait for the software to load. After the program is fully loaded, the ACCEL screen will appear and you will be able to choose between either live monitor and tuning or the rework of a file program.

Stop and Think

The EFI system is usually assumed to be the culprit of every drivability issue, but more often than not it is an innocent bystander. Before attacking a problem during a calibration procedure, I like to apply the SAT logic: stop and think. From my many years of experience diagnosing performance and drivability problems, I have learned that the time spent thinking and analyzing—and not jumping to conclusions and arbitrarily replacing parts or altering the calibration—pays huge dividends. Go through the following questions before putting either a wrench or a PC to the engine:

1) Was any work or maintenance performed before the problem occurred?
2) Have you looked for cracked vacuum hoses, disconnected spark plug wires, etc.?
3) Have you identified the exact symptoms and the engine load or operating rpm at which the problem occurs?
4) Is the condition the result of a mechanical failure or a poor choice of parts (too large a cam profile, incorrect fuel injector size, etc.)?
5) Have you listened to the exhaust to determine if the problem is dedicated to one cylinder or common to all?

If the problem is still elusive, then you need to determine whether the cause is fuel, ignition, or mechanical. I found the quickest method to determine this is to add fuel, subtract fuel, advance timing, retard timing. If the air/fuel ratio is altered and the timing advanced and retarded to no avail, then the poor performance is caused by something other than spark advance or fuel delivery. An ignition scope analysis, compression, leak-down, and power balance test is then in order. The best method to alter the fuel curve temporarily is to use the warm-up enrichment and air temperature correction tables. To add fuel, enter the warm-up enrichment screen and, at the operating temperature where the problem occurs, add five to 10 percent enrichment. To lean the mixture, use the air temperature correction table and subtract fuel at the temperature where the issue occurs. This method is quick and easy and does not alter the values in the base fuel matrix. Altering the phase of the distributor or changing the idle spark function can be used to experiment if spark timing helps the situation.

A good preventive maintenance schedule is usually the best method to avoid problems. Regular oil changes, fuel filter and ignition component replacement, and keeping the induction track free of vacuum leaks goes a long way toward trouble-free motoring. Driving habits can also add longevity. If you limit the number of cold starts and idling time, and do not command full power until the engine oil and coolant are at operating temperature, less wear and tear will occur. Being astute and acknowledging any peculiar sounds or behavior from the engine will allow small problems to be remedied before they become major issues. A true engine aficionado is constantly listening and recognizing the performance of his vehicle.

While in any screen, the user-friendly prompt of Q allows you to exit from that screen. Single-key quick prompts are available for access to most screens.

All program and tuning screens can be stored on the hard drive or disk for either future reference of to download into another ECU.

The CalMap software is extremely user-friendly but still requires a knowledge of engine and fuel-injection system functions. The best method to learn the software is to spend some time working with it, without any concern of doing an actual calibration. Familiarize yourself with the different screens and their corresponding submenus. Practice how to save a file and then retrieve it. Learn the quick prompts and how to execute changes. The more time spent using the software in the beginning, the easier it will be to learn how to tune an engine. After you master the execution of CalMap, you will realize that even though it is a very powerful tool, it is extremely easy to use.

Calibration Screens

The following is a screen-by-screen review of the CalMap software and a description of each screen's function. In this chapter I will cover the execution and function of each screen. The tuning process will be discussed at length in Chapter 8.

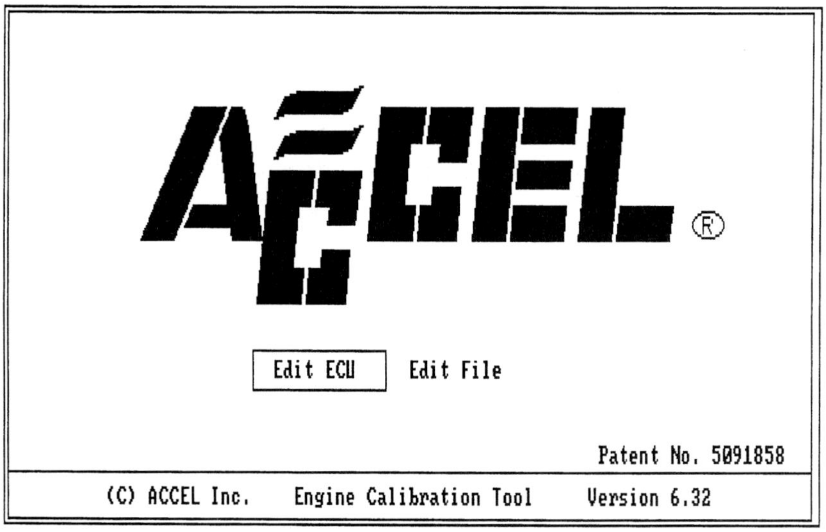

Figure 7-1

ACCEL Screen (Figure 7-1)

Once the CalMap program has been loaded, this screen will appear. On the bottom of the monitor will be a copyright statement and software version number. Above that will be two choices, with the cursor naturally returning to the EDIT ECU selection. To edit an existing file that has already been saved, use the arrow key to move the cursor to EDIT FILE and press enter.

To edit the ECU, the interface cable must be attached to the main wiring harness and the ignition on to supply power to the ECU. If the ignition is not turned on, or if there is a faulty connection at the main wiring harness, then a prompt will appear on the screen identifying a communication error. If the ECU is not powered up, then "Waiting for ECU response" prompt will be visible, along with a beep and a graphic display of a car going across the screen.

To edit a file after the prompt is selected, CalMap will ask for a configuration code. Enter 015013 for Spark/Fuel ECUs and 014013 for the Power Processor. In offline editing, a majority—but not all—of the calibration tables can be accessed.

After a selection is made, the main menu will appear. During offline editing, the Global function, which sets the base parameters, will not be accessible. All other main menu screens will be available, except for a few of the submenu functions. To choose a screen, simply move the cursor to the desired selection and press ENTER. To escape back to the ACCEL screen, press Q.

Base Fuel Map (Figure 7-2)

When the base fuel map is selected, you will have two choices: the core fuel table or a graph selection. The base fuel map selection is the actual programmable injector pulse table; the graph display is a contour-type graph that is used to visually spot any glaring errors in the fuel curve.

The fuel map is also identified as a matrix and consists of 240 set points: 16 rows across the horizontal portion and 15 vertical rows. The matrix is laid out with the horizontal representing rpm, and the vertical representing engine load. Load can be measured in bar or percent of TPS for alpha-numeric configurations. As stated previously, this is a look-up table, and the ECU will determine a place value by taking a reading for load and engine speed, and then assigning a place in the matrix for an injector pulse width.

When in live monitor mode, this screen will appear slightly different from the way it does in offline editing. The matrix values will be the same,

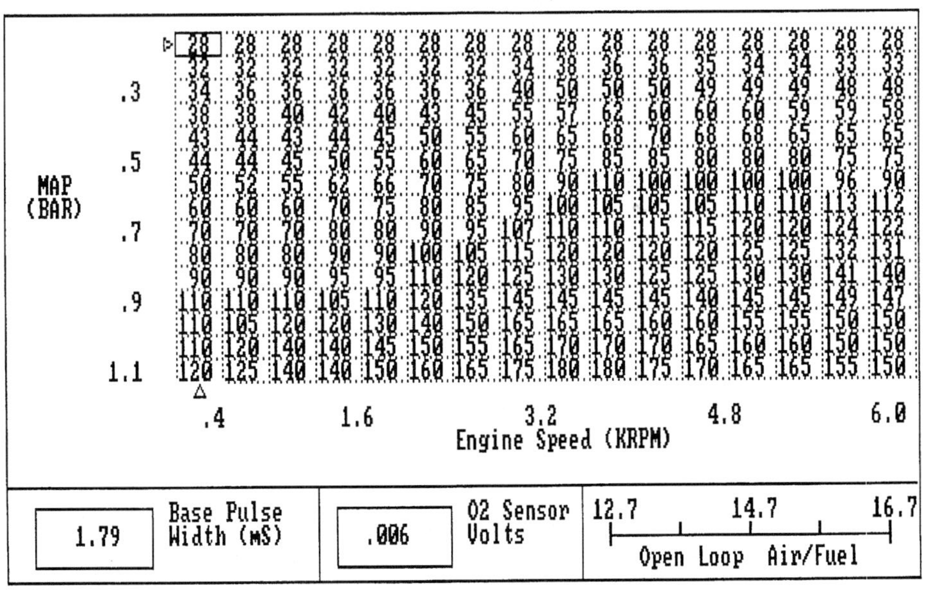

Figure 7-2

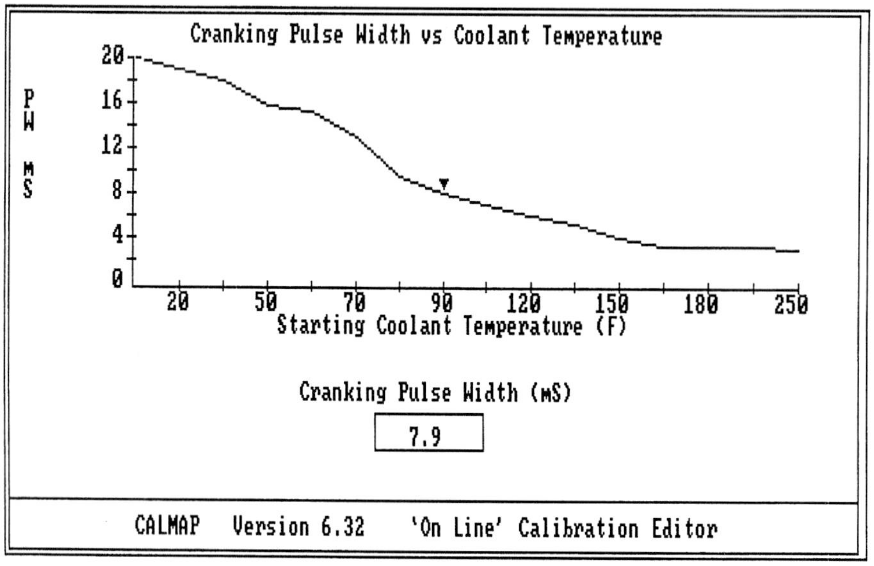

Figure 7-3

but during offline editing the load will be scaled in percent regardless of the ECU configuration. Also, the open-loop air/fuel monitor will be replaced by an rpm load table that identifies where the cursor is residing. During actual online editing, and if the ECU is configured for closed-loop fuel control, there will be an air/fuel ratio bar graph at the bottom of the screen. This graph has the ability to read two air/fuel ratios rich or lean of stoichiometric (12.7:1 to 16.7:1). The reading represents what the mixture would be if not corrected to 14.7:1. For example, if the air/fuel ratio graph read 13.5:1 and the ECU was configured in open loop, the injector pulse width in that position in the matrix would equate to a mixture of 13.5:1. Keep in mind, though, that the ECU will correct to 14.7:1, and this scale should be used to determine how much correction the ECU is evoking to achieve stoichiometric. If the ECU is configured for open loop, then the bar graph is replaced by a coolant temperature reading.

All the values in the matrix are numbers that, when multiplied by 0.0627, equal the commanded injector pulse width. For example, a matrix value of 119: The injector pulse width would be calculated as 119 x 0.0627 = 7.4613 ms. Since CalMap incorporates an interpolative and rounding function in its software, do not be alarmed if the mathematically calculated injector pulse width is slightly different from the valve assigned by the ECU. The software will take values from surrounding cells and blend them into the commanded injector pulse width.

The resolution of the load and rpm functions of the matrix can be altered to fit the tuning needs of a particular application. The rpm per cell division can be varied from 350 to 650 rpm, while the load section can be configured for one-, two-, or three-bar. Keep in mind that the amount of set points cannot be changed, just the steps between them. There are still only 240 places to work with regardless of the rpm and load configuration. Altering the scale in the matrix is covered during the explanation of the Global menu.

During live monitor there is a cursor that will identify in which matrix cell the ECU is operating for the current conditions. To change a programmed injector pulse width, press the E key (for Edit) and then scroll the cursor to the desired place in the matrix and type in the new value. Press ENTER to install the change.

Cranking Pulse Width (Figure 7-3)

This menu is used to program the commanded injector pulse width during cranking conditions and works independent of the base fuel map values. The screen is configured as a graph with the coolant temperature on the bottom and the injector pulse width on the vertical scale. There is a box-shaped cursor that identifies where the value is at that moment. Below 400 rpm, this is the injector pulse width that is used to start the engine and is not cumulative with the base fuel map matrix values. Once the 400-rpm threshold is surpassed, the values in the base fuel map become relevant. During this mode of operation, the programmed injector pulse width in this table is issued in an asynchronous mode, meaning that there are four injector pulses of the programmed value for each crankshaft rotation compared to one pulse per crankshaft rotation during running.

By toggling the arrow keys, a set point can be defined and a new value typed and entered by pressing the ENTER key.

Idle Speed

Submenu: Idle Speed—Sharing the same basic configuration as the cranking fuel table, the idle speed

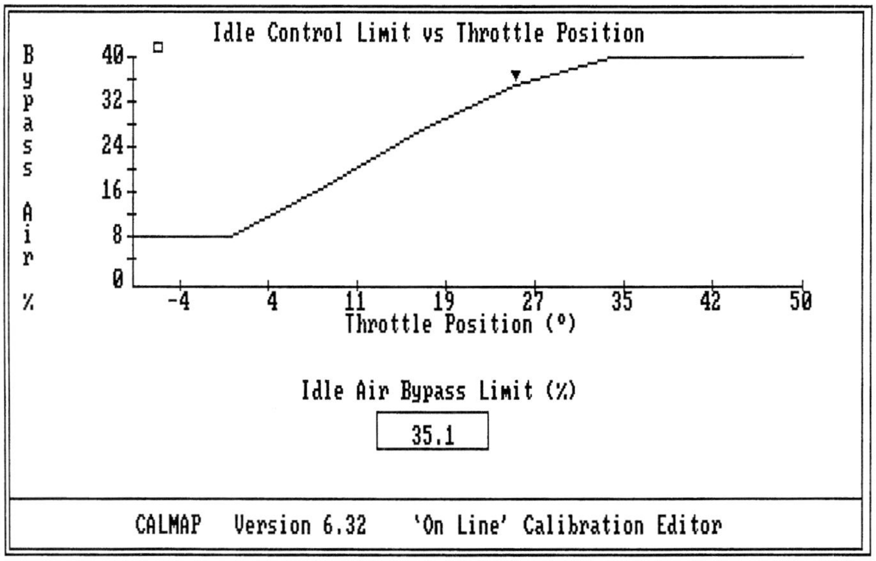

Figure 7-4

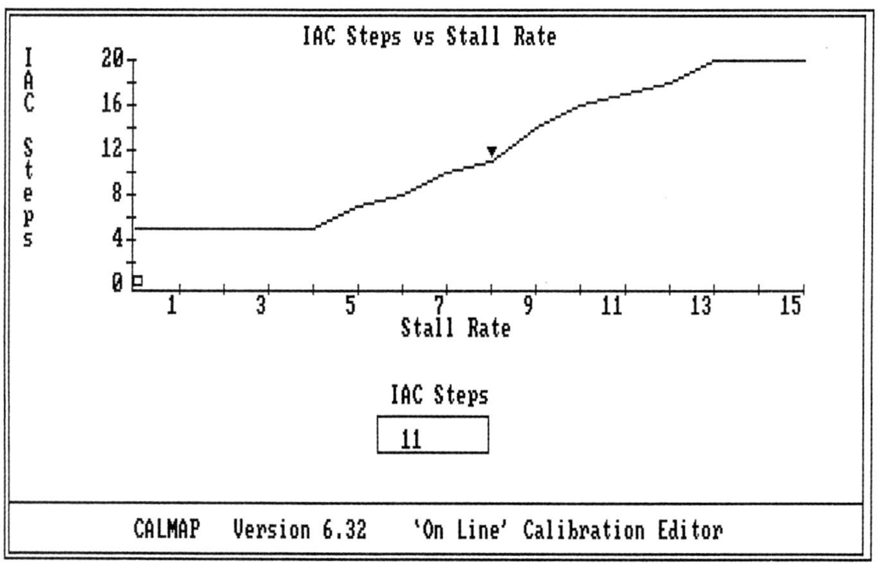

Figure 7-5

program is altered in the same method. The only difference is that the vertical side of the graph now represents a commanded idle speed for the given coolant temperature. At the bottom of the screen is a box that reads the programmed engine idle speed. The floating cursor box represents the actual engine idle speed when running and has the ability to travel below or above the set point value if that is where the engine speed is. The programmed idle speed can be accomplished only if the minimum throttle angle is sufficient to allow control through the idle air control motor.

Submenu: Throttle Follower (Figure 7-4)—Sharing the same layout as the previously mentioned tables, this represents the tracking of the idle air control motor in relation to the throttle angle. The vertical axis is the amount of air bypassing the throttle plates through the IAC passage, whereas the horizontal axis is the throttle angle read by the TPS. A floating cursor box identifies the actual amount of bypass air at that moment. This screen is used to set the minimum air rate by adjusting the air bypass screw on the throttle body and watching the cursor box. The function of throttle angle vs. bypass air acts similar to a dashpot on a carburetor, dampening the rpm degradation of the engine with a rapidly closed throttle. As with the other screens, the actual bypass air in relation to the arrow position is read in the box on the bottom of the screen.

Submenu: Stall Saver (Figure 7-5)—The vertical designation of this table represents the amount of IAC steps commanded to a programmed amount of rpm degradation represented on the horizontal axis. The stall values are arbitrary numbers that are assigned to a coded rate of engine speed drop. This table is used to compensate for a load dump on the engine that drastically affects idle rpm. This is used to compensate for a load such as shifting into a drive gear with an automatic transmission and a tight torque converter, or the load of an air conditioner compressor engaging. Then again, the customary floating cursor box and set point arrow work in conjunction with the actual place value box on the bottom of the screen.

Submenu: IAC Starting Position—As the name of this menu implies, it defines the IAC position on engine startup as a function of coolant temperature. The larger the number inputted, the more the IAC will be

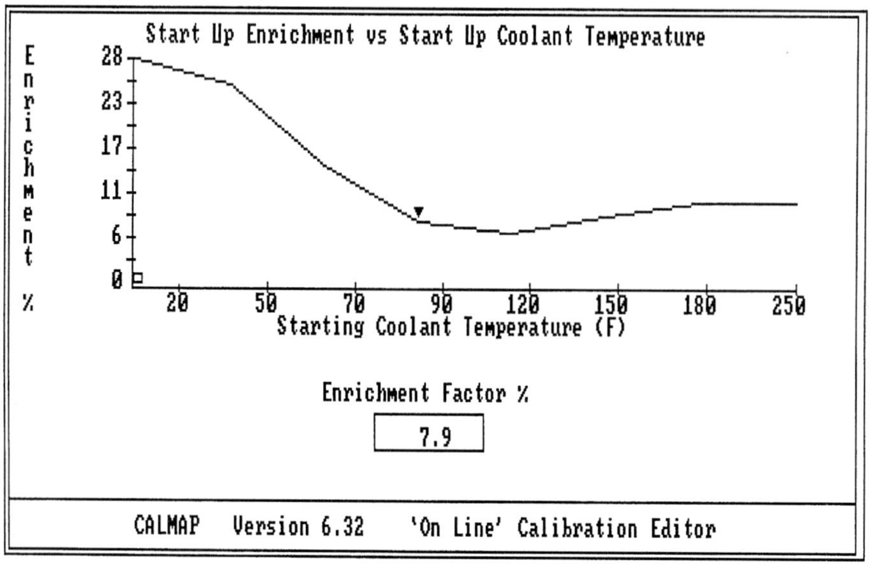

Figure 7-6

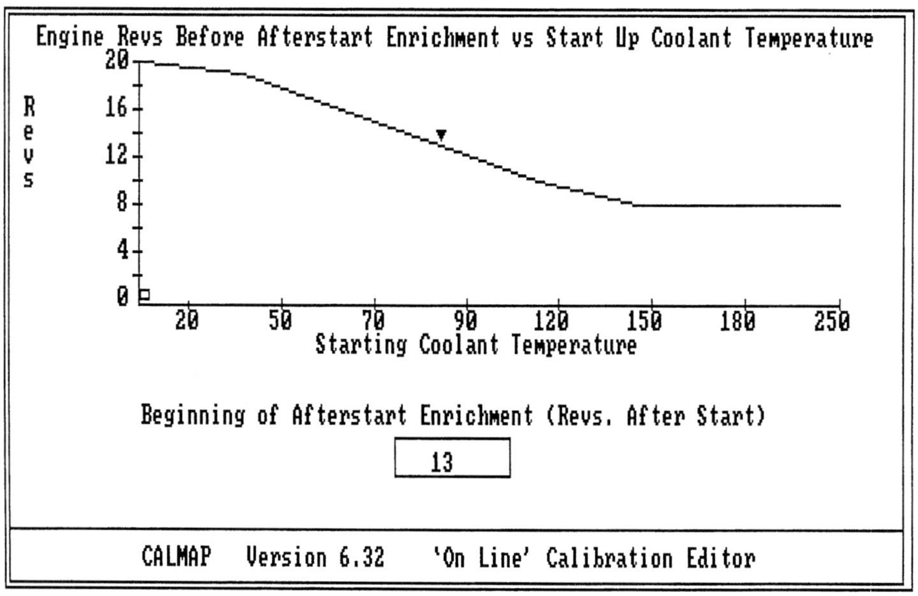

Figure 7-7

commanded to move to allow for a greater volume of bypass air.

Afterstart Enrichment

Submenu: Initial Afterstart Enrichment (Figure 7-6)—The afterstart tables are the hardest to understand in the CalMap software. They are the functions that control the mixture during the transition period from cranking to running fuel. They are also used to help keep the engine running during the transition from asynchronous to synchronous injection pulses. Unlike the cranking fuel table, which works independent of the base fuel map, the afterstart enrichment is an additive factor based on the values in the fuel matrix. Once the engine speed reaches 400 rpm, it is considered to be running and the cranking fuel is extinguished. As a transition to warm-up enrichment, which acts as a choke function, is where the afterstart values come into play. Taking the standard format as described for the other submenu tables, coolant temperature is scaled against a percentage of enrichment over the base fuel map value. Another function of this mode is to help in hot fuel handling during heat-soak conditions where it is common to boil the fuel in the injector rail.

Submenu: Beginning of Afterstart Enrichment (Figure 7-7)—Since the previous submenu defined the amount of enrichment that will be added, this table is the trigger for evoking the start of the enrichment. It is keyed from the number of revolutions of the engine after the cranking fuel set point threshold is crossed, in relation to the coolant temperature. All aspects of this table in regard to the floating box, arrow, and display hold true.

Submenu: Afterstart Enrichment Decay Rate (Figure 7-8)—Just the opposite of the previous table, this graph allows the user to set the rate at which the afterstart enrichment will be removed in relation to coolant temperature and revolutions of the crankshaft. Unlike the start of the additional fuel, the decay rate is controlled in a coded step rate that is triggered by a programmable amount of revolutions that need to be met. For example, if the value is set at five revolutions at 90 degrees F. coolant temperature, that would equate to one

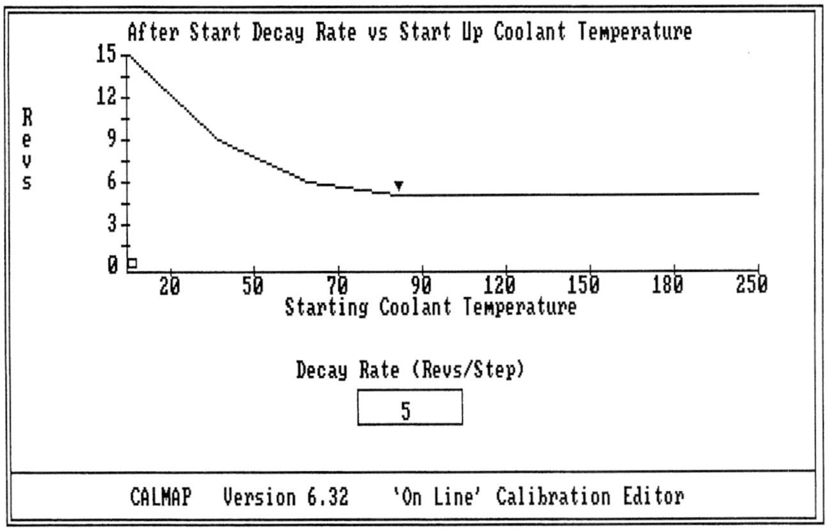

Figure 7-8

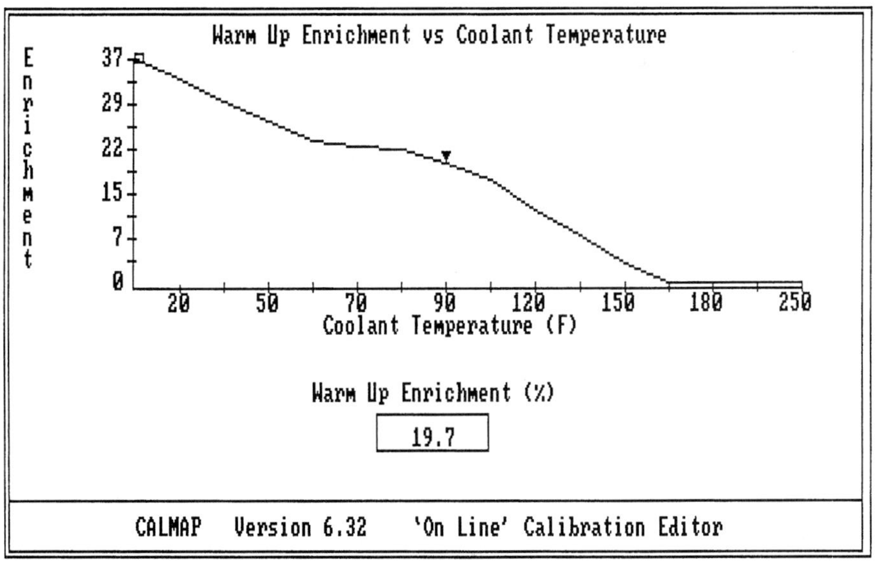

Figure 7-9

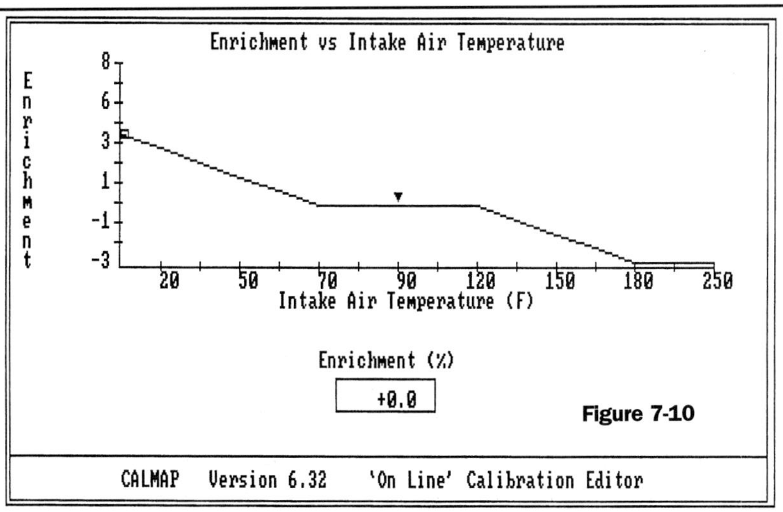

Figure 7-10

coded step of degradation for every five revolutions of the crankshaft.

Warm-Up Enrichment (Figure 7-9)

Once the transition from afterstart fuel is complete, the function of warm-up enrichment takes affect. It is analogous to the setting of the choke spring tension on a carburetor and the rate at which it opens. Scaled in the same format as the other submenus, it compares engine coolant temperature as a function of percentage of enrichment over the base fuel map values.

Air Temperature Correction (Figure 7-10)

When air is heated, its density decreases and it travels more slowly through the intake manifold, so it is essential to trim the base fuel map values to correct for this. A good rule is that for every 10 degrees F. the charge air temperature is altered, engine power is changed by one percent. As the incoming charge air is heated, the engine output drops, and the mixture goes richer if not corrected. Conversely, as the air is cooled, the air/fuel ratio is leaned if not compensated, and the potential for additional power is present. This table lets you add fuel for cooler temperatures and lean the air/fuel ratio when the charge is heated. The floating cursor box, arrow, and identifying box function the same as in previous screens.

Acceleration Enrichment

Submenu: TPS Acceleration Fuel (Figure 7-11)—This table controls the asynchronous injector pulse width, which is defined as the additional fuel pulses that are issued out of phase with the primary ignition interrupts.

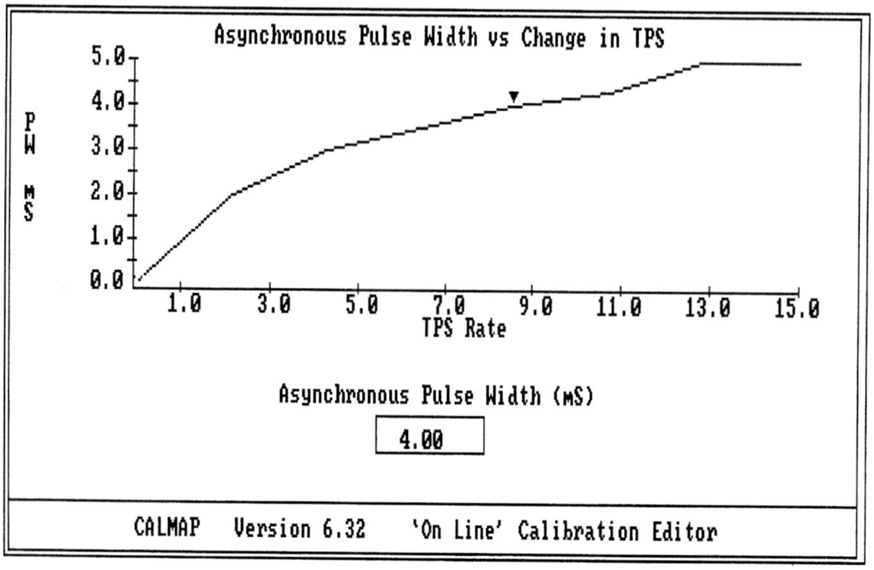

Figure 7-11

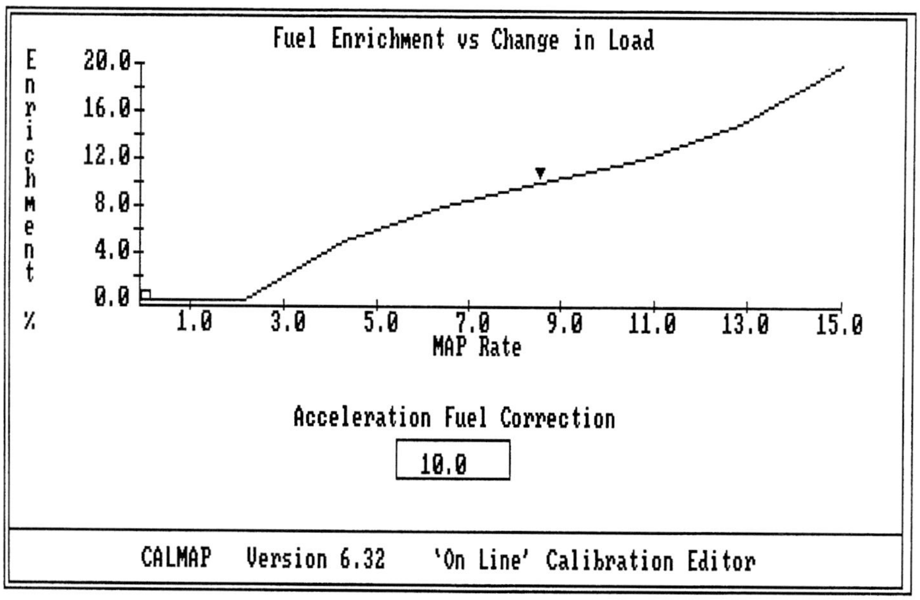

Figure 7-12

During asynchronous operation, the injectors are opened independently and in between each primary ignition interrupt triggered pulse. In this mode there are more injector pulses than primary ignition pulses. It is used as an accelerator pump in a carburetor and to administer additional fuel to compensate for a rapid opening of the throttle plates. The frequency or amount of additional pulses is coded into the software and is not user-adjustable. What is programmable is the amount of pulse width vs. an arbitrary rate of throttle angle change. This table uses the same graph format as the other submenus. Changes can be readily made and the asynchronous pulse width viewed.

Submenu: MAP Acceleration Fuel (Figure 7-12)—The MAP-evoked enrichment is analogous to the power valve in a carburetor and is used for enrichment to compensate for pressure changes in the intake manifold. It is triggered from an arbitrary rate of pressure drop in the intake manifold coded into the ECU as a percentage of enrichment from the base fuel map. Its primary function is to eliminate the need to over-fuel areas of the base fuel map to cover lean spots as engine load is transitioned while the vehicle is being operated. The layout and method of making changes mimics the previously mentioned configuration.

Ignition

When entering the ignition map function, you will have two options: a timing matrix that resembles the fuel matrix but is comprised of 64 points, eight horizontal for rpm and eight vertical keyed to engine load; or a three-dimensional plot of the timing matrix.

Submenu: Ignition Map (Figure 7-13)—The ignition map resembles a fuel map in appearance but with different scaling. Since the fuel map is larger, it has more resolution, which translates into the timing matrix being scaled at a rate of 1.5 between the cell sites. The rpm scaling of the fuel map then impacts the scale on the timing matrix. The following example will establish this concepts: If the fuel map was scaled with 400-rpm cell divisions, then the timing map would be scaled at 1.5 x 400, or 600-rpm cell width.

Below the matrix is a boxed value that reads the actual timing command that the ECU is issuing. As with the fuel map, this value interpolates from the surrounding cells and may not agree with the value at which the cursor is residing. In live monitor mode, an additional boxed display will be available representing the amount of knock retard that is taking place to control abnormal combustion, if the ECU is equipped with the optional knock retard kit. The knock sensor will remove timing at the rate of 100 degrees per second with a maximum of 20 degrees. It then listens for knock and advances the timing at a rate of one degree per second.

Submenu: Graph Ignition Map—This function is used to provide a contour graph display of the programmed timing curve so that any discrepancies or error in tuning will become quickly apparent. This function can be accessed only during online editing.

Note: All of the above menus can be saved in a file independent of the other menus. For example, if you wanted to save only the idle table from a particular calibration, it can be stored in the appropriate submenu file. All files that are saved as submenus can be accessed only during online communication.

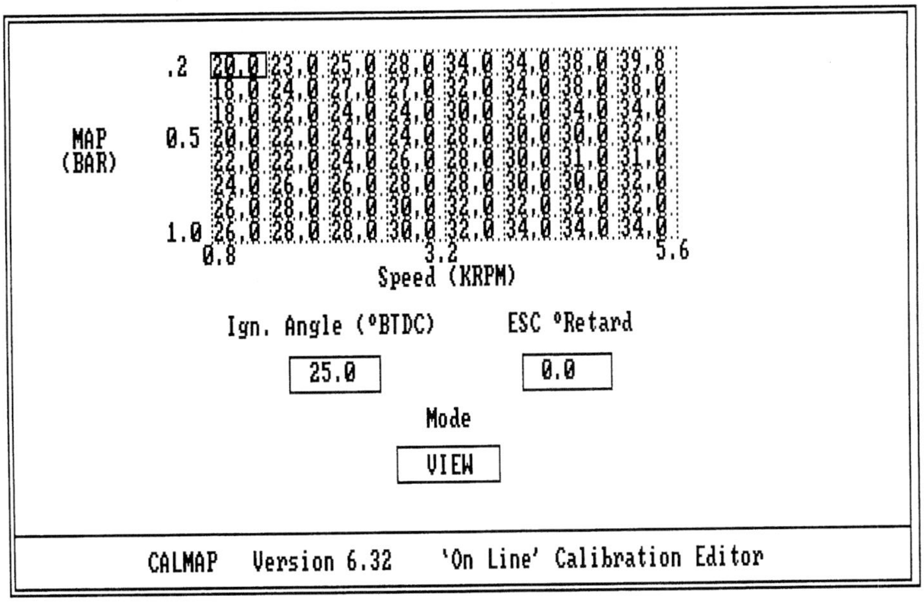

Figure 7-13

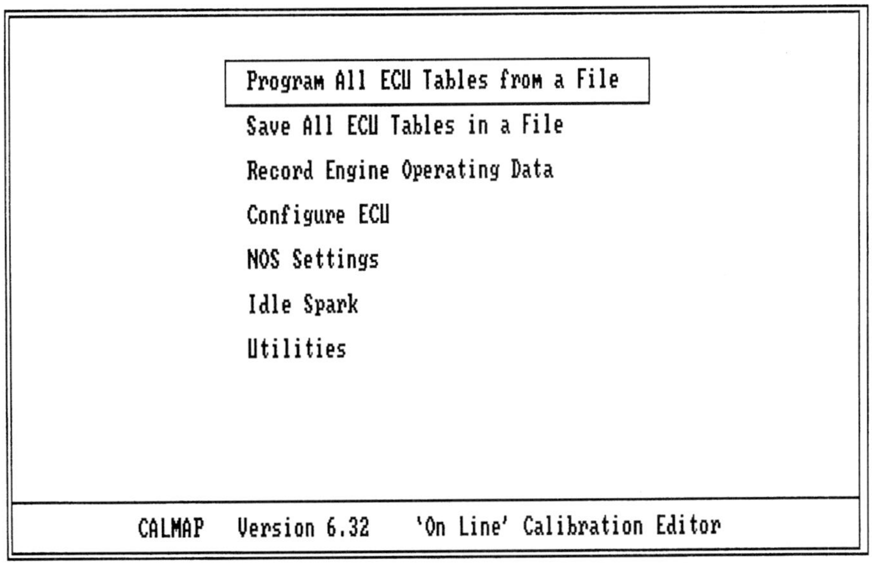

Figure 7-14

Global

When selecting from the root menu, a Global submenu listing of seven choices will appear.

Program All ECU Tables from a File (Figure 7-14))

If executed, this will rewrite all of the ECU tables from a program that was already saved on the disk from a previous calibration. All aspects of the chosen program will be inputted into the ECU that you are currently connected to. If the program from the ECU that you are working on is not saved prior to this routine, it will be lost.

Save All ECU Tables in a File

This will allow you to save the complete calibration of the ECU in a named file for future use. This routine should be executed first if you are going to reprogram the entire calibration from a saved file, affording the ability to revert to the original program if desired.

Please note that these submenus pertain to the complete calibration, all of the tables and files. Tables that were saved as individual files in submenus while performing online editing will not be accessed. For example, if you want to use a previously created calibration in a different vehicle, then reprogram the entire ECU from a file. But if you want to alter just the timing table from

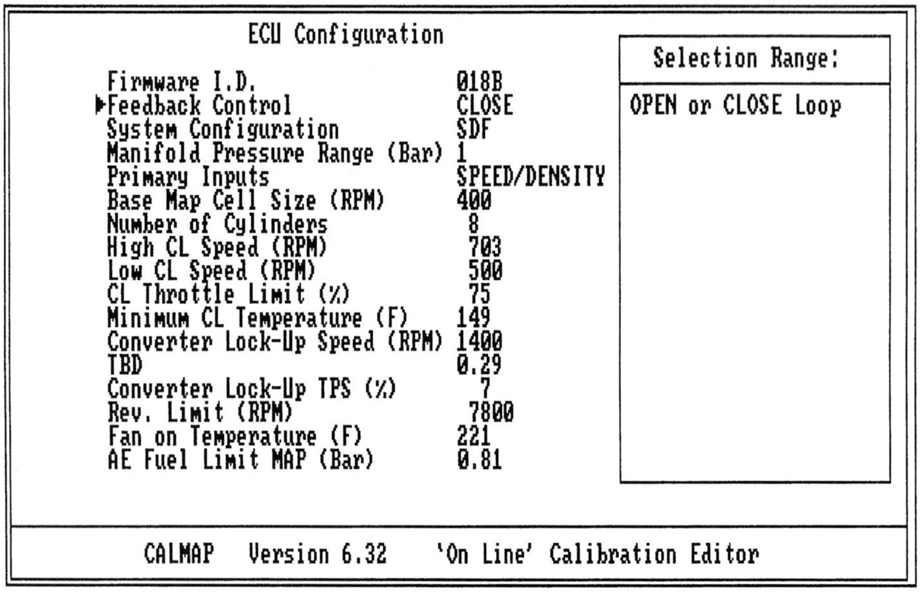

Figure 7-15

a different ECU, then after completing the Global function of reprogramming, you will need to go to online editing and reprogram the timing table from the ignition submenu.

Record Engine Operating Data

When selected, the CalMap soft-ware will record up to six user-chosen data points and store them on a disk for review. While this is being logged, you will be able to see the data just as the ECU witnesses it. This function is designed to be a diagnostic tool and tuning aid. Reference Chapter 6 for execution.

Configure ECU (Figure 7-15)

Used to configure the ECU for base parameter selection and to inform the software of the engine that it is operating.

The following is a list of configurations and their descriptions:

Firmware ID—The Firmware ID is what CalMap uses to distinguish between ECU types and can not be altered.

Feedback Control—This is either open or closed loop.

System Configuration—Simultaneous double-fire (SDF) is used for port fuel-injection applications, or TBI for throttle body injection. A TBI system for a V-8 engine has the injectors opened once every two ignition events; however, the SDF triggers the injectors once every fourth ignition firing.

Manifold Pressure Range—There are three ranges offered: one-, two-, and three-bar. A one-bar sensor has the ability to read only vacuum. Two- and three-bar sensors register from 15 inches of mercury to 15 or 30 psi of boost pressure, respectively. This information is used to scale the load scale of both the fuel and timing matrix.

Primary Inputs—This controls the reference for load on the Y axis of the fuel and matrix. Either speed/density, which is manifold absolute pressure, or alpha-numeric, which is throttle angle, can be chosen.

Base Map Cell Size—This is designed to alter the cell width of the x axis of the fuel map and ranges from 300 to 650 rpm per cell division. It allows the user to tailor the resolution for the specific application.

Number of Cylinders—Used in parallel with the system configuration to set the correct injector firing routine. Choices are four, six, or eight cylinders.

High-Speed, Closed-Loop—The engine rpm that must be exceeded before switching to closed-loop operation. This works in tandem with the low-speed, closed-loop setting.

Low-Speed, Closed-Loop—Sets the low-engine rpm threshold hysteresis for the feedback operating range. High- and low-speed closed loop are used to allow an engine to idle in open loop and then enter closed-loop operation after the high-rpm setting is surpassed.

CL Throttle Limit—The throttle limit expressed in percent, when exceeded, will revert to open-loop operation. This is required to avoid having the ECU drive the mixture to 14.7:1 during WOT operation. Due to the nature of CalMap, even though this setting has the ability to be at 100 percent, when reading TPS in the record data mode or on the view screen, the display will show a maximum value of only 99.9 percent.

Minimum CL Temperature—Expressed in degrees Fahrenheit, this is a parameter that must be met to operate in closed loop. This prevents closed-loop operation during cold engine operation.

Converter Lockup Speed (rpm)—A threshold value that must be reached to engage the torque converter clutch when using a lockup kit.

Converter Lockup MAP—Replaced by the fourth gear discrete signal of the GM 700R4 Turbo-

Hydramatic transmission. As a result, this function is no longer active.

Converter Lockup TPS—The throttle position expressed in percent of opening that must be exceeded for the torque converter clutch to disengage.

Rev Limit—The engine rpm a fuel shut off will be evoked to control engine speed. There is a 1,000-rpm hysteresis coded into this function, which requires that amount of rpm degradation before the fuel is turned back on.

Fan-On Temperature—Displayed in degrees Fahrenheit when the electric cooling fan will be turned on. When this occurs, the idle speed will automatically be increased by 50 rpm to compensate for the load on the charging circuit. There is a coded 10-degree hysteresis that needs to be met to turn the fan off. Regardless of the fan-on temperature, at WOT the ECU will automatically open the fan control circuit.

AE Fuel Limit (Bar)—In Speed/Density configuration, this value will be the limit for activating the MAP-based acceleration fueling. In simpler terms, this is the point that, when surpassed, there will no longer be any MAP acceleration enrichment. It is analogous to the point at which the power valve in a carburetor will no longer give any additional enrichment. When configured for alpha-n operation, it represents the TPS percentage, even though the software will still read in bar.

Nitrous Settings—The standard 015013 ECU has the ability to control one stage of nitrous oxide and the corresponding tuning changes that need to be made for proper and safe operation. Through this function, the nitrous oxide engagement throttle position, time delay, rpm setting, fuel enrichment, and timing retard can be controlled.

Idle Spark—The purpose of the idle spark function is to add an additional measure of idle speed control through the use of ignition timing. Large-capacitance intake manifolds are very sensitive to reversion from camshaft overlap and the inherent idle instability that it causes. ACCEL has found that varying the amount of timing in accordance with the amount of idle instability works very well to control this problem. To enter the idle spark function on software prior to version 6.32, you must be online and choose EDIT FILE on the ACCEL screen. Then type in the word IDLESPARK instead of a configuration code and press ENTER. That will display a screen that allows changes to the amount of spark advance vs. idle rpm variation. It functions by changing the timing command on the ECU pin P2-B1. The idle spark function is engaged only above 100 degrees F engine coolant and below 1.5 percent TPS.

Back to Basics

Before considering any major investment in performance parts, the engine should be confirmed as a worthy candidate. All too often huge sums of money and time are spent on a mill that has poor ring seal, worn bearings, and a sloppy timing chain. To get the most from your performance dollar, do the following tests before you buy any expensive item:

1) Compression test: Reveals the engine's ability to pump air and compress the charge.
2) Leak-down test: Confirms the cylinder's ability to seal during combustion. Worn piston rings and valves, along with the condition of the bore, will be determined.
3) Timing chain slack test: There is no use trying to tune an engine that has an excessively worn timing chain. As the chain wears, camshaft timing will retard from the intake centerline.
4) Mechanical oil pressure test: Connect a mechanical oil pressure gauge and determine the health of the bearings and clearances. There should be at least 10 psi of oil pressure for every 1,000 rpm.
5) Oil analysis: A laboratory oil analysis costs only $20 or $30 and can be done by one of the many companies across the country. You can search the Internet under "oil analysis" or speak to the manager of a local truck stop or trucking or equipment fleet for information in your area.

8

Tuning with CalMap Software

Now that you have a basic familiarization of the ACCEL ECU and its companion components, the intricacies of the system can be discussed. As stated in the beginning of the book, to tune this system you need a working knowledge of OE fuel injection. This will make it much easier to master the ACCEL design.

As mentioned in Chapter 7, the first step should be to become familiar with the software and not concern yourself with making any tuning decisions. If you try to learn the software and tune the engine at the same time, you'll not only drive yourself crazy, but you'll also risk damaging the engine. Once you feel comfortable accessing the different screens and understand their purposes, the tuning of engines can begin.

When working with a programmable ECU, your startup calibration may be way off. It will be up to you to determine if it is a calibration error, mechanical problem, or both. With a stock ECU, the only variables are the engine and its ancillary parts, not the program. That is why ACCEL has invested the time to develop many different calibrations for the Chevrolet manifold and camshaft combinations that are popular. By ordering the proper part number ECU, you will bring yourself closer to the mark. As your tuning career grows, you will quickly see the benefit of establishing your own library of calibrations for different engine combinations.

Where to Begin

Most problems have their start with poor preparatory work. Either the operator is not familiar with basic engine and fuel-injection functions, or he did not go back to basics in checking over the engine prior to the tuning session. From my experience, I

CALCULATING STATIC FLOW
Static Flow = 56,000/rpm
Horsepower = 5252 x Horsepower/rpm
BSFC = Fuel pounds per hour/brake horsepower

POWER LOSS IN COMPARISON TO SEA LEVEL
BHP Loss = Elevation in feet/1000 x 0.03 x Bhp at sea level
Theoretical cfm = rpm x displacement/3456
Volumetric Efficiency = Actual cfm/theoretical cfm x 100

CALCULATING INJECTOR SIZE
Lbs./hr. = Max hp x bsfc/number of injectors

The proper way to perform a calibration is by doing the math first. Use these formulas to help you determine key factors.

believe that there are a few distinct scenarios when tuning a spark/fuel ECU.

One is an engine, ECU, or both that were not installed or built at your shop. This is not to say that the engine, fuel injection, or components were installed improperly, but that you have a good chance of finding something wrong. You start tuning and are trying to compensate for a vacuum leak, a faulty ignition, a poorly assembled engine, a faulty ECU installation, poor atomization by the injectors, or a totally wrong combination.

The second case is that you do the installation of the ECU but the engine is an unknown variable. Another variation on this is that you install the injection manifold on an unknown engine.

The best scenario is that you build the engine and install and tune the fuel-injection system. Then you can choose the proper components to obtain the desired results.

Quite often you will find that the combination of components, especially the camshaft and intake manifold, is wrong for that application. Most enthusiasts and engine builders pick a camshaft or cylinder head that would work with a carbureted engine and a short-runner intake manifold. It is also common to fall into the drag race syndrome of "bigger is better" and install too large a fuel injector. That is why this book establishes the basics first.

Before doing any tuning on a vehicle that you're unfamiliar with, get as much information from the owner as possible. Ask about the compression ratio, camshaft specifications and installed position, fuel system and components, and basic ignition condition.

Perform a visual inspection of the engine and test drive the vehicle if possible. Also do a scope analysis, looking for possible problem areas. It is very common to have vacuum leaks with a poorly installed GM TPI. With the engine running, go around the

Injector Flow in Lbs-Hr. vs. Fuel Pressure

Injector Size @ 3 Bar	Fuel Pressure 40	45	50	55	60
19 lbs-hr.	18.21	18.32	20.37	21.38	22.31
22 lbs-hr.	21.09	22.37	23.58	24.73	25.83
24 lbs-hr.	23.01	24.41	25.73	26.98	26.18
30 lbs-hr.	28.76	30.51	32.16	33.73	35.23
36 lbs-hr.	34.52	36.61	38.59	40.47	42.27
55 lbs-hr.	52.74	55.94	58.96	61.84	64.59
83 lbs-hr.	79.59	84.41	88.98	93.32	97.47

Use this handy chart to determine changes in flow rates for popular injector sizes.

8 CYLINDER HORSEPOWER vs. BSFC vs. FUEL PRESSURE

FUEL PRESSURE	BSFC .42	.45	.47	.50	.55
19 LB/HR INJECTOR					
40	346.85	323.73	309.95	291.36	264.87
45	368.00	343.46	328.85	309.60	281.01
50	388.00	362.13	346.72	325.92	296.29
55	406.85	379.73	363.57	341.76	310.69
60	424.95	396.62	379.74	356.96	324.50
22 LB/HR INJECTOR					
40	401.71	374.93	358.97	337.44	306.76
45	426.09	397.68	380.76	357.92	325.38
50	449.14	419.20	401.36	377.28	342.98
55	471.04	439.64	420.93	395.68	359.70
60	492.00	459.20	439.65	413.28	375.70
24 LB/HR INJECTOR					
40	438.28	409.60	391.65	368.16	334.69
45	464.95	433.95	415.48	390.56	355.05
50	490.09	457.92	437.95	412.16	374.25
55	513.90	479.64	459.23	431.68	392.43
60	536.76	500.97	479.65	450.88	409.89
30 LB/HR INJECTOR					
40	547.80	511.28	489.53	460.16	418.32
45	581.14	542.40	519.31	488.16	443.78
50	612.57	571.73	547.40	514.56	467.78
55	642.47	599.64	574.12	514.56	467.78
60	671.04	626.84	599.65	563.68	512.43
36 LB/HR INJECTOR					
40	657.52	613.68	587.57	552.32	502.10
45	697.73	650.84	623.14	585.76	532.50
50	735.04	686.04	656.85	617.44	561.30
55	769.52	719.46	688.85	647.52	588.65
60	805.14	751.46	719.48	676.32	614.83
55 LB/HR INJECTOR					
40	1004.57	937.60	897.70	843.84	767.12
45	1065.52	994.48	952.17	895.04	813.67
50	1123.04	1048.17	1003.57	943.36	857.60
55	1177.90	1099.37	1052.59	989.44	899.49
60	1230.28	1148.26	1099.40	1033.40	939.49
83 LB/HR INJECTOR					
40	1516.00	1414.93	1354.72	1273.44	1157.67
45	1607.80	1500.62	1436.76	1350.56	1227.78
50	1694.85	1581.86	1514.55	1423.68	1294.25
55	1777.52	1659.02	1588.42	1493.12	1357.38
60	1856.57	1732.80	1659.06	1559.52	1417.74

intake manifold and runners checking for leaks with a propane enrichment tool. Repair any problems before tuning. With a used intake manifold, make sure the throttle body and IAC are clean and carbon-free. Check the fuel pressure to confirm 43 to 45 psi at atmosphere (vacuum hose removed from the fuel regulator and the engine running). Perform a thorough inspection on the car injector cleaning to ensure good cylinder-to-cylinder fuel distribution. It would also be advisable to zero the idle spark. It does a very good job at masking ignition, air leak, and injector problems.

Base Adjustments

Once you're confident that the installation of the fuel injection system and the engine are sound, you can make the base adjustments in preparation for doing a calibration. These are the prescribed steps:

1) Configure ECU

The first part of any calibration is to check the configuration of the ECU. Make sure the configuration matches the engine in terms of primary inputs,

Most drag race cars are configured to run in alpha-n.

Injector Static Chart For SDF

RPM	PW	Matrix Value
3500	17.14	273
3750	16.00	255
4000	15.00	239
4250	14.12	225
4500	13.33	212
4750	12.63	217
5000	12.00	191
5250	11.42	182
5500	10.90	173
5750	10.43	166
6000	10.00	159
6250	9.60	153
6500	9.23	147
6750	8.88	141
7000	8.57	136
7250	8.28	132
7500	8.00	127
7750	7.74	123
8000	7.50	119

Note: For SEFI, ECU's use the same matrix values to identify static opening. But due to the frequency difference in the firings of the injectors in SDF vs. SEFI, the indicated gross pulse width in the VIEW screen will be double the matrix value.

number of cylinders, etc. It should be application-specific to the engine you are tuning.

Here are some suggestions for configurations with explanations of their logic.

Feedback Control—The prescribed method of tuning is to utilize a wide-band air/fuel ratio monitor to determine the mixture. This lets you complete the calibration faster and more accurately. If an independent monitor is going to be used, then configure this mode for open loop, since the ACCEL-supplied oxygen sensor will be replaced with the one from the meter. If a stand-alone air/fuel meter is not available, then configure the ECU for closed loop and utilize the air/fuel monitor on the base fuel map screen.

System Configuration—This choice is straightforward: SDF or TBI, whichever is applicable.

Manifold Pressure Range—Normally aspirated engines will always use the 1-bar setting. If the vehicle is supercharged or turbocharged, examine the amount of boost that will be created. In theory, whenever forced induction is being used, a 2- or even 3-bar sensor needs to be employed. The problem is that each time the manifold pressure input scale is raised, the resolution of the fuel and timing maps is decreased. A 2-bar sensor will cut the resolution in half, and a 3-bar will reduce it to 1/3 compared to a normally aspirated calibration. With every reduction in resolution, you are forced to try to tune the engine with much less fuel and timing control. For supercharged applications that are running below 10 psi of boost for short periods of time, use a 1-bar sensor. A forced-induction application that will spend most of its

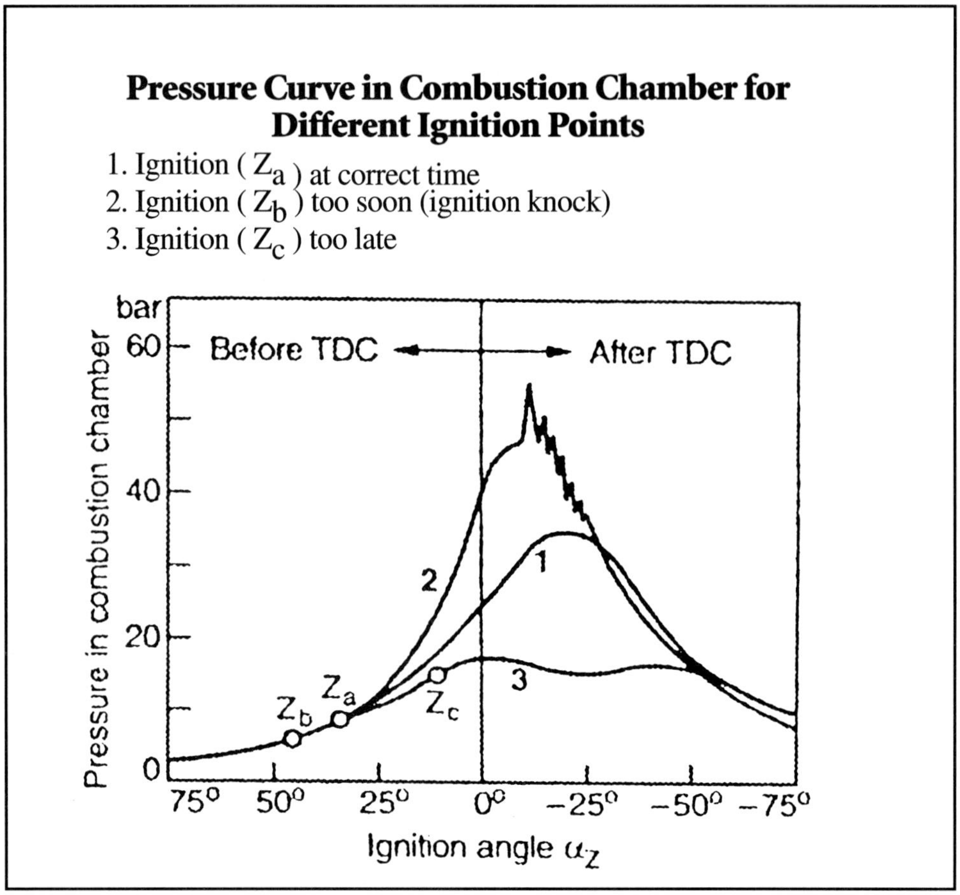

This chart represents a cylinder pressure curve for varied spark strategies.

time in boost (race engine) will necessitate the use a 2–bar sensor. For almost all street-driven turbos up to approximately 24 lbs. of boost, use a 2-bar sensor for the best driveability. Pressures above that will require a 3-bar sensor. The logic is that most supercharged street engines spend the majority of time in vacuum, or zero vacuum, with boost only produced for short periods at higher rpm. Since a turbo spools and builds boost independently of rpm, the time spent in boost is usually greater than with a supercharger, and therefore would require more resolution under boost.

Primary Inputs—For most applications, choose speed/density. All forced-induction systems must use speed/density. If you are working with an engine with very little or no vacuum (below five inches of mercury), then there will be hardly any resolution in a speed/density-based fuel map. For this reason, it is better to use alpha-n and key the load calculations from throttle angle input. Whenever switching from speed/density to alpha-n, the vacuum hose must be removed from the MAP sensor and only a 1-bar sensor can be used. There is no direct correlation from a speed/density to an alpha-n map; all of the fuel and spark tables will need to be redone.

Base Map Cell Size—This function will control the amount of rpm resolution the base fuel and timing maps have. The tendency is to want to plot the map out to a high rpm. Again, think about how the vehicle is going to be driven and the engine that is in it. With a long-runner manifold (TPI or Super Ram), peak power will occur between 4500 and 5000 rpm. By using a map cell size of 300 to 350 rpm, the scale will end at 4800 to 5600 rpm, respectively. Even if the engine makes power to 6000 rpm, it can still use the 350-rpm scale. When the engine speed goes beyond the scale, the last value in the matrix is read and used for timing and fuel control. For example, if the map is scaled to 4800 rpm with the injector pulse width for that cell being 9 ms, then the ECU will continue to issue a 9-ms pulse above that rpm point. This allows the maximum number of tuning set points at light-to-intermediate load, producing the best driveability, best throttle response, and cleanest emissions.

Number of Cylinders—This is self-explanatory.

High and Low Closed-loop Speeds—I have placed these together since they are dependent on each other. When configured to operate in open loop, these values serve no function but will have to be inputted if closed-loop fuel control is desired. When operating in closed loop you will have to decide if you want the engine to idle in closed or open loop. If closed-loop idle is desired, set both of these values below the idle speed with a 100-rpm difference. One reason for not having closed-loop idle fuel control is the camshaft in the engine. Historically, any degree of overlap will require a richer-than-stoichiometric mixture at idle. This can be accomplished by setting the low rpm value 100 rpm below the base idle speed, and the high rpm value 100 rpm above the base idle speed. In between those values, the ECU will revert to open loop. If emissions readings are a concern and a catalytic converter is being used,

Optimized electronic ignition-advance map (top) compared with the ignition-advance map of a mechanical spark-advance system (below)

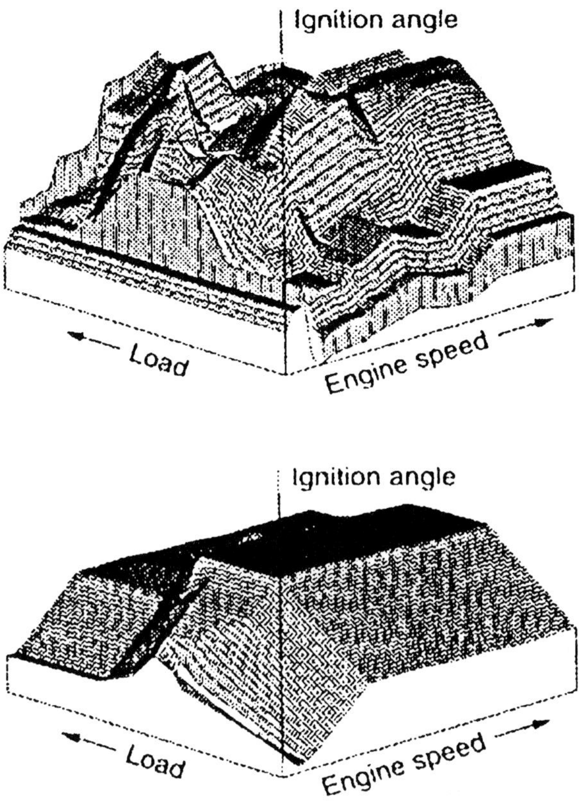

The bottom graph represents the timing curve with a conventional mechanical advance distributor. The top is what the engine wanted for best performance, obtained with electronic timing control.

conversion efficiency ramps off considerably when the mixture drifts more than 2 percent richer or leaner than 14.7:1. For this reason, it may be desirable to idle in closed loop. When configured for closed-loop idle, a heated oxygen sensor is required to maintain accuracy, especially with headers.

CL Throttle Limit—Once the closed-loop speed threshold is surpassed, the ECU will stay in feedback mode until the throttle angle input is reached. If you are not using a stand-alone air/fuel meter, then, while tuning, you can consider setting this value at 100 percent. This will allow the use of the open-loop air/fuel meter at WOT as a tuning aid. Do not do sustained WOT passes with this function set at 100 percent. If a stand-alone air/fuel meter is being used, this setting is moot until the system is put back in closed loop. For normal operation when tuning is complete, a value of 70–80 percent is desirable.

Minimum Closed-Loop Temperature—This represents the temperature that needs to be reached for closed-loop operation. In most applications it is best to set this at approximately 150 to 160 degrees F. Make sure that the setting coincides with the decay of the warm-up enrichment. Ideally, the warm-up enrichment should be at zero 20 degrees prior to this threshold.

Converter Lockup rpm—The converter lockup function references only rpm and throttle angle. An OE system also looks at coolant temperature and, more important, road speed. This ACCEL function does not always give desirable results at lower rpm. Try different settings while noting the rpm at which the vehicle cruises at highway speeds.

Converter Lockup MAP or TBD—This is not used. Any changes here have no impact.

Converter Lockup TPS—Functions in conjunction with the rpm setting and is usually sensitive to each installation. A good starting point is 20 percent.

Rev Limiter—This is a fuel shutoff with a 1000-rpm hysteresis. Set it near the point that you choose to limit the engine speed. If the vehicle has an ignition rpm limiter, the ACCEL ECU will not have the ability to override the setting if adjusted at a lower speed value.

Fan-On Temperature—It is usually best to set this value 10 to 20 degrees above the thermostat setting so that the fan does not run constantly. There is a ten-degree hysteresis that must be met to shut the fan off again.

AE Fuel Limit (Bar)—This is the operating state referenced to manifold pressure at which MAP fuel enrichment will cease to be evoked. A

The cranking speed is critical for proper starting, especially on a high-compression engine.

An air filter on the throttle body may look cool, but it forces the engine to breathe hot air from the radiator.

good starting value on normally aspirated engines is .70, and on turbocharged and supercharged applications use .90. Use this setting to increase the crispness of the throttle at high loads. Lowering the value cancels enrichment sooner, whereas raising it holds it in longer. In alphanumeric, this configuration refers to TPS percentages, but the scale does not indicate this.

2) View Screen

After configuring the ECU and before starting the engine, the View function should be employed with the

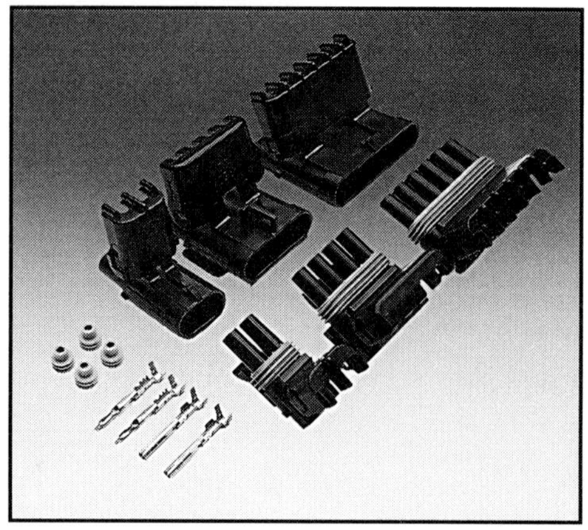

Use Packard-supplied ACCEL connectors for a weatherproof seal on any ECU circuit.

V key. In this mode, CalMap works like a scan tool and displays all of the sensor inputs and some of the ECU outputs. Use this to make sure the ECU is communicating with all of the sensors.

3) Review All Screens

By reviewing all screens one by one, you will have the opportunity to check the program before starting the vehicle. Pay attention to anything that would be detrimental to the engine, such as an overly rich calibration, excessive timing, or idle speeds.

4) Zero Idle Spark

The reason for negating the idle spark is to prevent it from masking any mechanical problems that might not have been discovered during your pre-start testing. The idle spark function was added to help maintain a constant idle speed, not as a crutch.

5) Save Program

Enter the *Global* function and save the entire program under *Save All ECU Tables in a File*, and identify it with a name that you will associate with this particular vehicle. Execute this routine in the early stages of your learning curve. If you get totally off

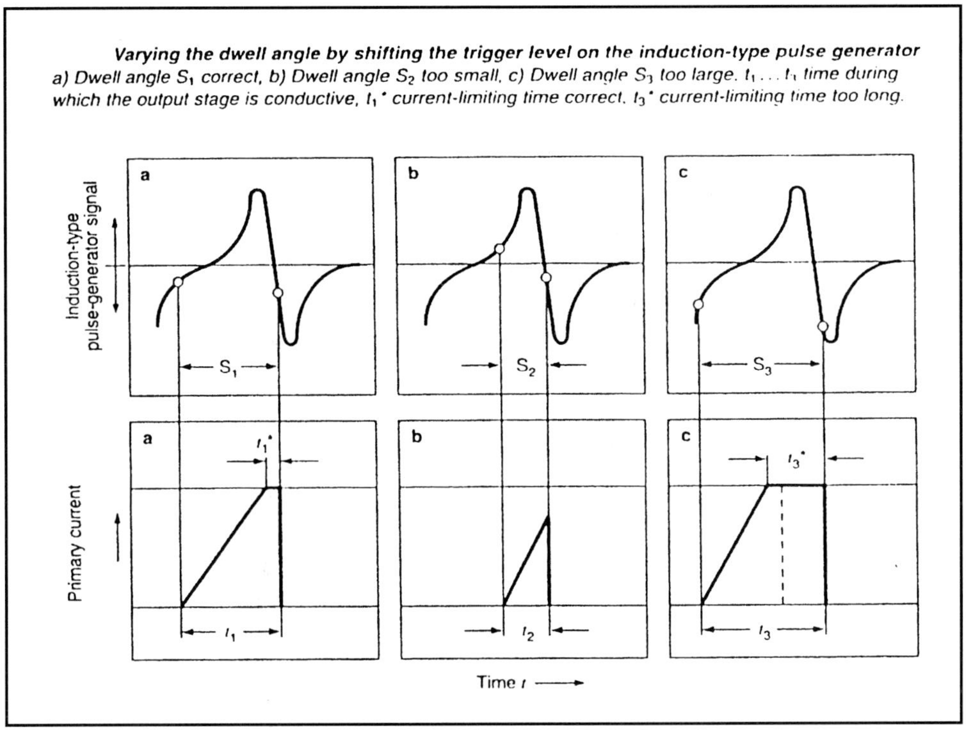

If the polarity is reversed on the IPU wires going to the ECU, the ignition timing will be retarded 20 degrees.

Sufficient charging circuit voltage is essential for proper ECU operation.

Often something as simple as a worn distributor gear will haunt a calibration, haunting the tuner.

base, you can always revert to the original calibration. As your experience and your library of files grow, you can bypass this step.

6) Start Engine

If the engine fails to run on the third try, stop and examine what is going on. Pull a spark plug and check for flooding. If the plugs are wet and the engine never ran during your attempts, go to cranking fuel and lean it out. If the engine started and then stalled within the first three to five seconds of running, the problem is most likely in the afterstart enrichment. Do not concern yourself with that now. Try restarting the engine. It will probably run.

If the opposite is happening when you examine a plug and you see no fuel, then richen the cranking fuel table. If the engine stays running but appears to be very rough, then it is most likely too rich. As long as it is not blowing black smoke, just let it warm up and the problem will be addressed later. Keep in mind that it does not take much to foul the plugs or wash the rings out of a port fuel-injected engine due to the close proximity of the injector to the intake valve. Allow the engine to warm up until the cooling fan cycles once, if so equipped.

7) Adjust TPS

Enter the *View* screen (V) and adjust the TPS to .5 percent. This will allow the ECU to recognize a closed throttle position. A small amount of preload is desirable since it guarantees the TPS will track any change in throttle movement.

8) Check Idle Speed

While in the View screen, check to see if the idle speed is in range. If it is, continue to the next step. If it is too high or low, enter the idle speed menu and check what rpm is commanded. If the IAC cannot accomplish the commanded idle speed, then the throttle plates are too far closed. If the actual idle speed is substantially over the commanded value, then the throttle plates are too far opened or there is an excessive air leak. Keep in

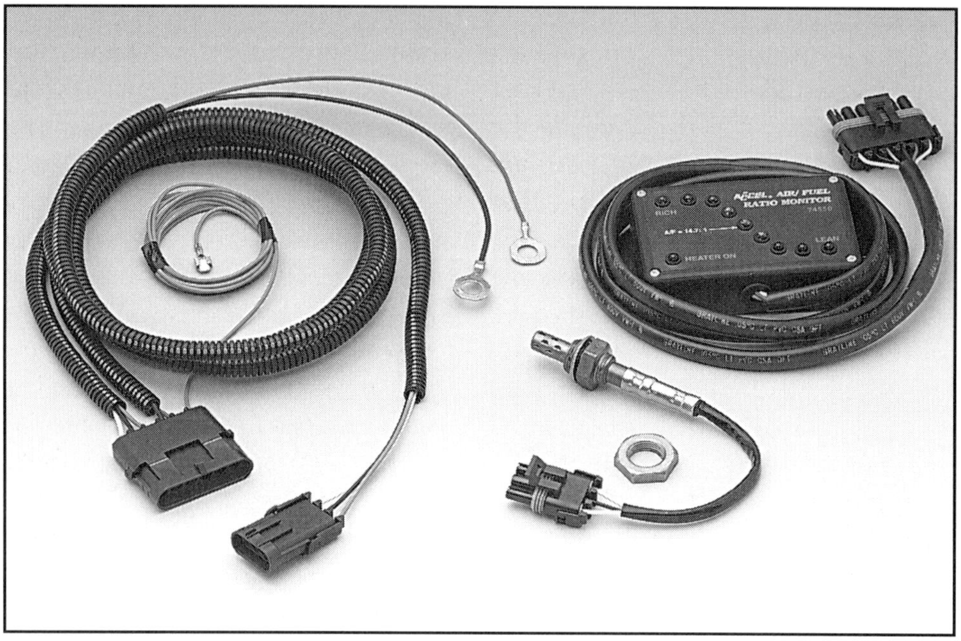

An in-car air-fuel meter is a good way to keep track of the mixture, especially on a nitrous or forced-induction engine.

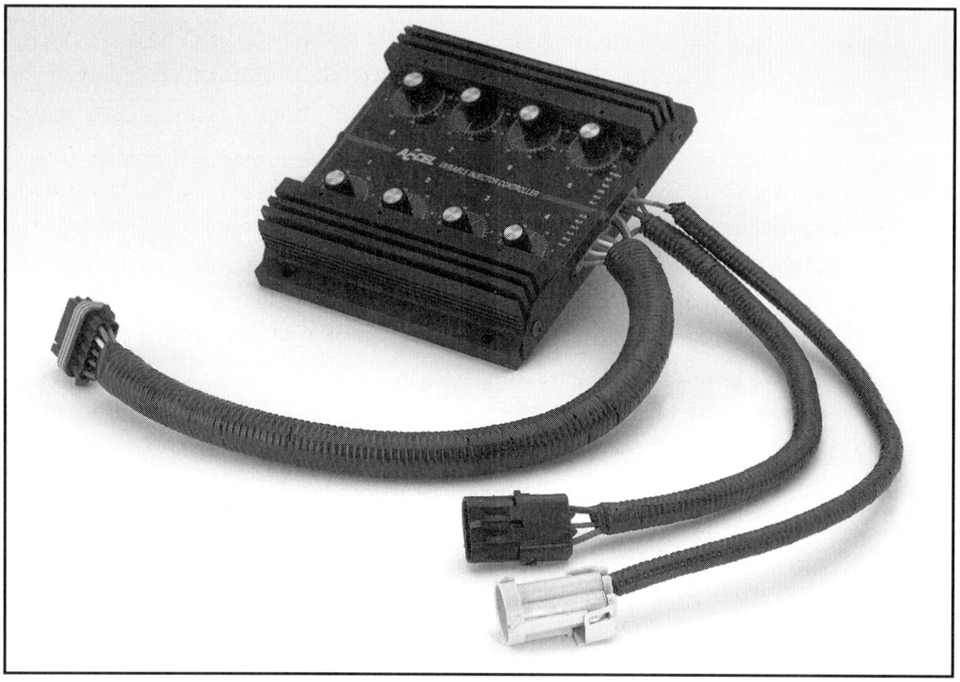

The installation of a VIC will allow individual cylinder tuning and quicker injector opening times.

mind that as of yet, the base timing was not set, so this table may need to be revisited later. To adjust the throttle plates, enter the *Throttle Follower* submenu in the idle speed tables and adjust the throttle stop screw to achieve approximately five to 10 percent bypass air. Since the throttle plate angle was altered, readjust the TPS. If the commanded idle speed is wrong, input the value you desire.

9) Set Timing

When setting the timing on an ACCEL ECU, it is critical to synchronize the reading on the harmonic balancer and the base timing matrix. If this is not done, the timing program may not be accurate. To synchronize the timing, first enter the ignition menu and save the current timing map in a subfile by executing *Save Ignition Map in a File*. This will allow you to revert to the original timing program without reprogramming the whole ECU. Once the map is saved and titled, return to the timing matrix and press the E key to place CalMap in *Edit* mode. Since the timing map interpolates values from one cell to another, it is six degrees for Chevy and 10 degrees for Ford. Whatever value you use, convert the whole matrix to that number. You will now have a map with all the same numbers. Start the engine and let it idle. With your timing light connected physically, adjust the distributor until the value on the dampener reads the same as the value in the boxed area below the timing matrix. Note that during this initial setup procedure, the idle spark function has already been zeroed. If you are just going to check the timing on a vehicle and not executing the complete base calibration, make sure you raise the TPS to over 1.5 percent to disable the idle spark function. It is easier to disable it this way instead of erasing an established idle spark program. For the initial tune-up, I choose to zero the idle spark since a calibration has not yet been established. When the timing procedure is complete, return the TPS to a setting of .5 percent and enter the already saved timing map by executing *Program The Ignition from a File* routine.

10) Drive Vehicle

At this point I like to drive the vehicle for a few miles to clear it out and get some heat into the oxygen sensor. Do not make any WOT passes, but you can drive aggressively to make sure the plugs are cleaned out. If the vehicle will not drive due to an improper fuel calibration (extreme bucking, hesitation, surging), do not try to force it. The fuel map will need to be reworked first.

11) Base Fuel Map (Idle)

Once you return, start your programming by working from idle to light load to intermediate load to WOT. Enter the Base Fuel Map screen by pressing B, or enter through the base fuel map routine. Locate the cursor to see the commanded injector pulse width. If you are using a stand-alone air/fuel meter or the CalMap open-loop scale, change the numbers in the matrix to achieve the best idle with the leanest possible air/fuel ratio. As stated previously, an engine equipped with a cam with some degree of overlap most likely will not want to idle at 14.7:1. If a satisfactory idle is unobtainable, then return to the *Global Configure ECU* menu and change the low and high closed-loop settings to idle in open loop. When working with the fuel matrix, always remember that CalMap is an interpolative software and the fuel cells around the cursor also affect the gross pulse width. It is common to have to make changes to the surrounding cells to achieve the proper mixture. An efficient way to dial in the fuel map is to look at the amount of correction needed to arrive at the desired mixture and alter the map value by that percentage.

For example, if you return to the *View* screen, there will be a listing for the amount of oxygen sensor correction. If this value is reading -12 percent and the cursor in the matrix is sitting at a value of 44, then subtract 12 from 100 and then multiply 44 by .88, to come up with 38.72. Round this value up to 39 and enter that number in the matrix. Then take the surrounding cells and make the same adjustment. To richen the matrix value by a percentage, just add the desired percentage to 100 and multiply by that number. To establish this, use the same equation. For example, to add 12 percent enrichment to a matrix value of 44, multiply 44 by 1.12, and you come up with 49.28. If it is determined that the complete fuel map needs to be altered, you can either lean or richen the calibration from its current values using the following steps:

1) Save entire calibration to a file.
2) Access the Base Fuel Map of the just-saved calibration. This needs to be done in the *Edit File* mode from the ACCEL screen.
3) Press M, A, or S to increase or decrease by a percentage (M), add a number of counts (A), or subtract a number of counts (S) to the entire map.
4) Now save the updated calibration to a file.
5) Enter *Program All ECU Tables from A File* using the new calibration name.

The updated calibration has now been loaded into the ECU. This is usually needed if the injectors are too large or small for the ECU configuration ordered. Keep in mind that when modifying the base fuel map by counts, it will have a greater effect on the low-load numbers then on the high-load values. It will most likely be necessary to alternate between the fuel map, timing map, and idle speed tables to achieve the best idle characteristics. Adjust for idle quality the way you would on a carbureted application, but instead of using a screwdriver, distributor wrench, and timing light, it can all be accomplished from your laptop.

If you are still unable to achieve a decent idle, it's time to evoke the idle spark function. To enter this screen on the older software, escape back to the ACCEL screen and enter *Edit File*. In lieu of entering a configuration code, type IDLESPARK (all capitals) and enter that menu. Start inputting an idle spark value for each rpm variation. Begin with zero for the least variation, and go to a maximum of six degrees for the largest variation. A value greater than six can be used but usually indicates a problem in the engine or elsewhere in the calibration. After the idle spark is programmed, return to the fuel map and work the idle cells for the best idle quality with the least emissions. It may be possible now to enter closed-loop idle control. Try it and see.

12) Idle Timing

Use the timing map to help achieve the best idle quality and throttle response. Keep in mind that, like the fuel map, the cells around the matrix value will affect the actual timing command. If the idle spark function is enabled, it is normal to see the actual timing command in the box below the matrix vary.

13) Base Fuel Map Light Load

Since the idle calibration is now very close to being done, start thinking about off-idle fueling. You may want to start by simply raising the rpm without placing a load on the

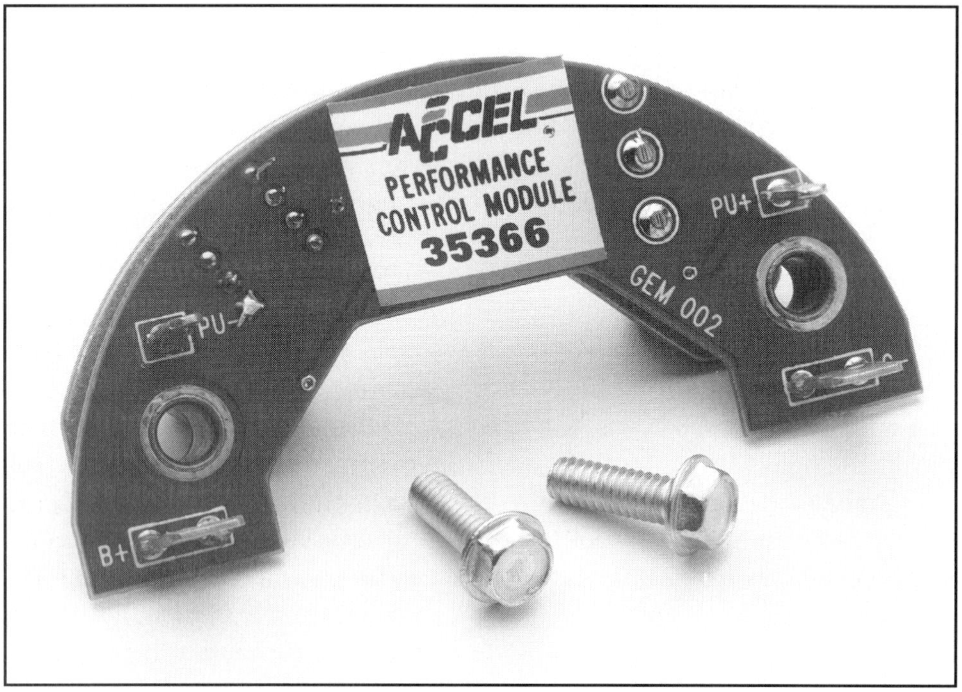

An ACCEL high-performance ignition module should be part of your EFI package.

engine, and calibrating the fuel cells the same way you did for idle. Do this to a maximum of 2000 rpm. Most engines do not require much change from the idle setting as the rpm increases and no load is present. Usually the top numbers of the fuel matrix will vary minimally from idle to 3000 or 4000 rpm. If the vehicle is equipped with an automatic transmission, it is desirable to chock the wheels and set the parking brake and do some light-throttle power-brake tuning to 1500 rpm, or the stall speed of the torque converter, whichever comes first.

Historically, the acceleration enrichment functions are near enough to being correct that, once the base fuel map is tuned, little if any adjustment is needed in these tables. To properly and safely tune a fuel map, you will need to bring along a driver or access a chassis dyno. Using the air/fuel readings at light load, try to obtain a mixture ratio near 14.7:1. In almost all instances, this is a desirable target. Follow the same procedure that you did for idle by using the V key to see how much oxygen sensor correction is needed, and use that value to modify the fuel matrix.

Tell your driver to start out very easily and work his way up to more aggressive driving patterns so that the look-up table may identify different parts of the fuel matrix. A knowledgeable tuner will try to emulate all types of driving conditions to make full use of the tuning capabilities of the ECU. If transitional fueling leaves huge holes to achieve the proper balance of driveability and air/fuel ratio, then the function of acceleration enrichment can be used to make the transition smoother. It is usually the most desirable to create a fuel map that, at light load, relies very little on the oxygen sensor to achieve stoichiometric. The closer your map is to being correct, the less correction from the oxygen sensor will be needed. If correction is to be made, it is usually desirable to have the oxygen sensor lean the calibration instead of richening it. This leaves an extra margin of safety against damaging the engine if the oxygen sensor circuit fails. Follow this procedure up to approximately 70 percent load.

14) Ignition Timing Part Throttle

Now that the base fuel map has been optimized up to 70 percent load, the timing map can be modified without concern about being lean in these areas. Since we have already synchronized the ECU with the distributor, all of the tuning can be accomplished with the laptop. If the subject vehicle is equipped with a knock sensor kit, then the task will be much simpler due to the sensor's ability to recognize detonation. Ideally, you would want to utilize as much spark lead as you can without evoking detonation. A "seat-of-the-pants" feel is usually the best tool for this type of application.

Keep in mind that the ECU has the ability to identify 64 set points for timing control. This differs from a conventional mechanical/vacuum advance distributor due to the ability to advance or retard the spark for the best performance during a specific load and rpm. The ECU has the ability to run more timing at idle, pull it back out at light load, add more under moderate load, and then adjust for WOT conditions. In most instances, the tuner does not recognize this ability and sets the timing curve as if it were a conventional distributor. Study the way the cursor travels the matrix for varying load conditions, and input timing commands to achieve the best driveability. It is customary to start with a very conservative timing curve and become more aggressive in small

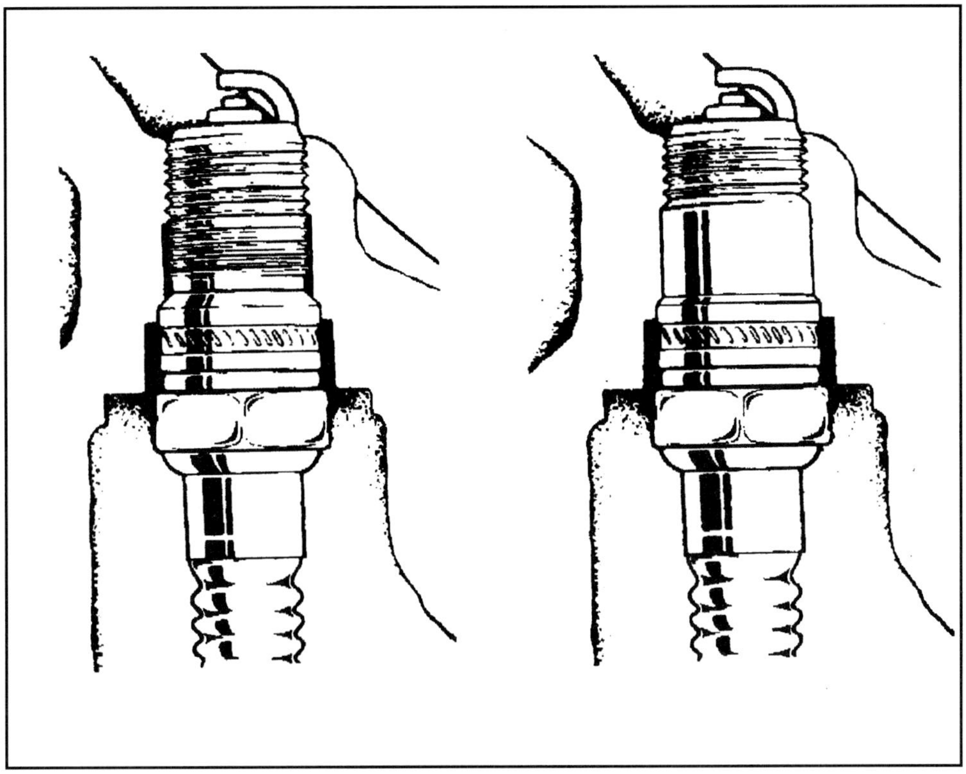

Extended-reach spark plugs that place the electrode closer to the center of the combustion chamber will require less spark advance.

increments until light detonation is detected.

15) Wide-Open Throttle Fuel Map

With the timing and fuel maps now calibrated up to the 70 percent load threshold, tuning for maximum power can be done. This is where a stand-alone air/fuel analyzer comes in handy. Most engines will want an air/fuel ratio between 12.7 and 12.9 at peak torque and lean out to 12.9 to 13.1 at peak horsepower. This is why, when viewing the *Graph Base Fuel Map* function, there will be longer injector on time at the torque peak. This segment of the tuning is best done on a chassis dyno due to its repeatability and the legal and moral issues of racing down the highway at WOT.

When using an air/fuel meter, make a wide-open-throttle pull watching the way the cursor travels the fuel matrix while acknowledging the air/fuel ratio readings. You will be able to identify the peak torque rpm if a dyno is not used by the engine's highest demand for fuel. To determine the needed injector pulse width, divide the actual by the desired air/fuel ratio and adjust the fuel map accordingly. For example, if the actual air/fuel ratio is 13.1 and you want 12.9, divide 13.1 by 12.9 to come up with 1.01 percent. Then richen the matrix in that area by 1 percent. If an air/fuel monitor is not available, you can use the open-loop air/fuel ratio monitor on the bottom of the base fuel map screen. This scale reads what the mixture would be if not driven back to 14.7:1. The problem with using this is the possibility of leaning the engine, since the mixture will be driven to 14.7:1.

On normally aspirated applications, short bursts at this setting usually will not hurt anything, but forced-induction engines are a complete different story. For this reason I don't suggest using this method for turbocharged or supercharged applications. Forced-induction engines usually want a slightly richer mixture to help cool the charge. It may be advisable to run these engines at 12.2 to 12.5 air/fuel ratio at peak torque, and 12.5 to 12.7 air/fuel ratio at the horsepower peak. It is better to give away a few horsepower to an

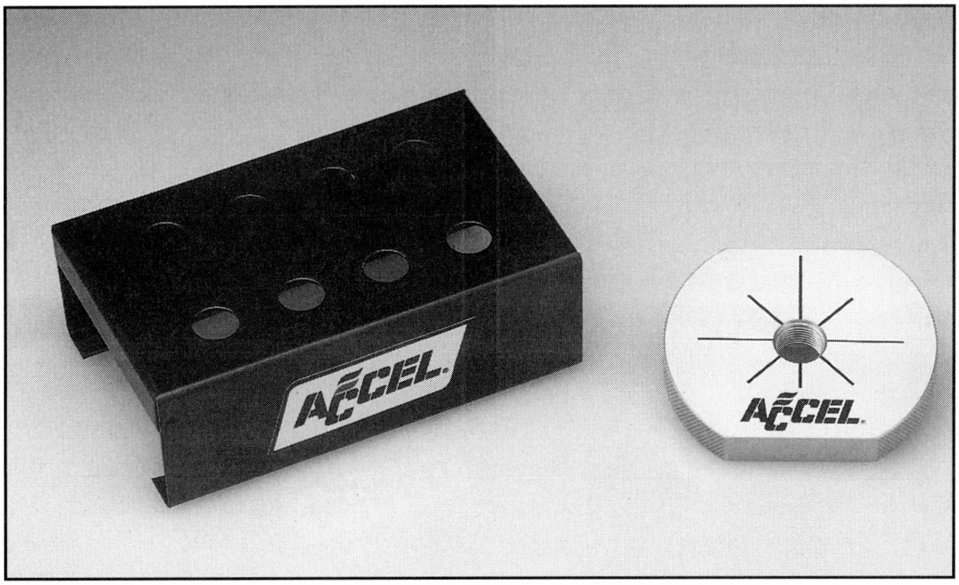

Older-style combustion chambers respond well to indexing of the spark plugs.

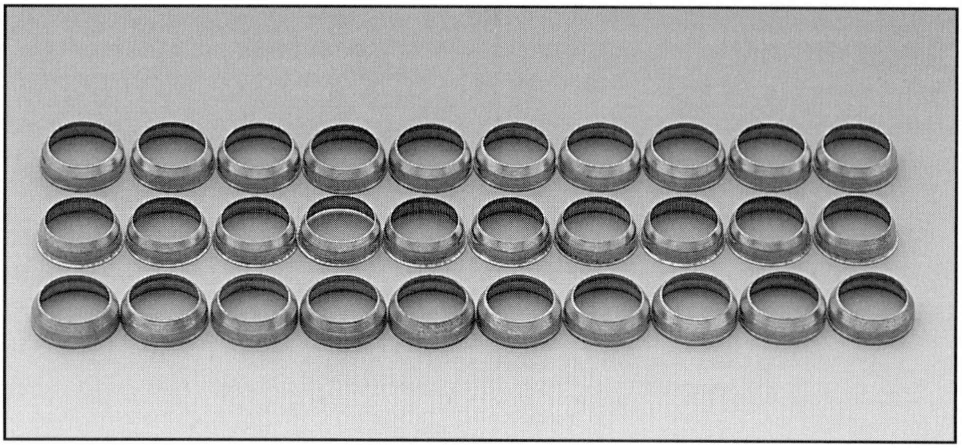

Index washers are used to adjust the clock orientation of the spark plug side electrode.

overly rich mixture than to blow a head gasket or lift a piston ring land. Always keep in mind that CalMap is an interpolative software, and the fuel cells around the actual cell where the cursor resides have an affect on determining the injector pulse width.

16) WOT timing

Now that the fuel curve is correct, the WOT ignition can be adjusted for the most power. Once again, this area of tuning is best done on a chassis dyno for safety and the ability to quantify the results. Prior to the advent of electronic timing controls, most engines had a compromised spark curve, derived from rpm reference only. The ability to adjust the timing was relatively crude. As stated previously, most tuners carry this same logic through to electronic spark timing, not optimizing the timing for different load conditions. An important fact to recognize about ignition when using a closed-loop timing control kit is the theory of the hysteresis of knock.

For example, if maximum brake torque is achieved at 30 degrees BTDC, and the timing is advanced to 31 degrees BTDC and abnormal combustion occurs, it may be necessary for the ESC system to retard the timing by as much as 10 degrees to eliminate knock. Keep in mind that once abnormal combustion begins, it takes a disproportionate amount of timing retard to stop the detonation. This is due to the super heating of the combustion chamber and the need to excessively retard the ignition event to cool down the region. For this reason, the amount of knock retard that is displayed cannot be used to determine the amount of error, unlike the oxygen sensor correction with the base fuel map. Always start with a conservative timing map and work toward a more aggressive program.

17) Cold-Start Tables, IAC Starting Position, and Cold Idle Speed

At this point the vehicle should be completely tuned, and the only issues that may be unresolved are in afterstart fueling and IAC control. If the program is plagued with a cold-start stall, then the *After Start Enrichment* needs to be adjusted for a lean condition. This usually reveals itself after a true cold start, which is defined as the first startup of the day. This condition may be very elusive and temperature-dependent, making it hard to calibrate. To expedite the correction, it is advisable to make drastic changes in this table, essentially over-correcting.

Identify the coolant temperature at which the problem occurs and work in that area. Conversely, if the engine stays running after a start but acts as if it is very rich by running rough (similar to a carburetor with a faulty choke pull-off) the afterstart

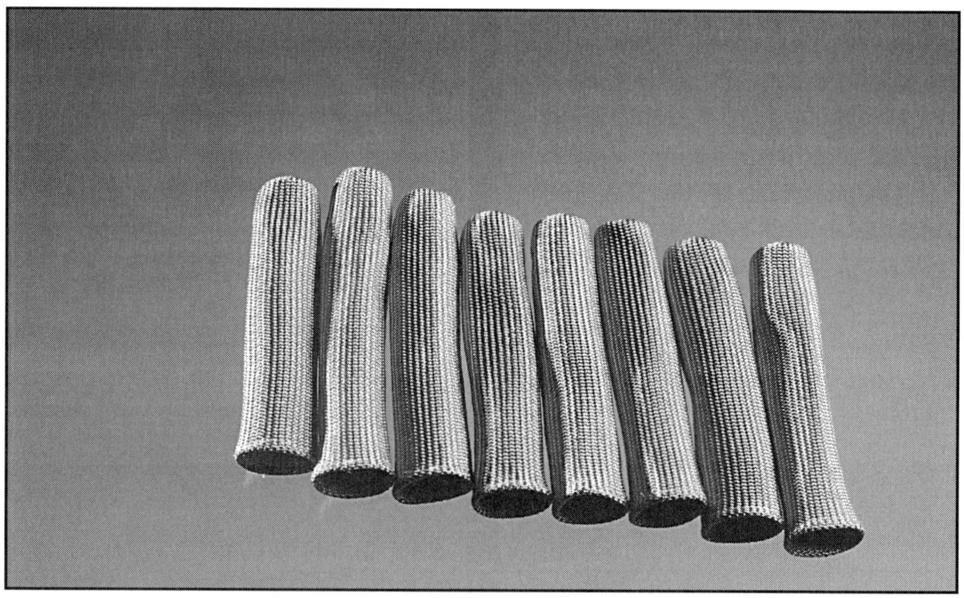

High-heat-resistant spark plug boots are required on most turbocharged engines.

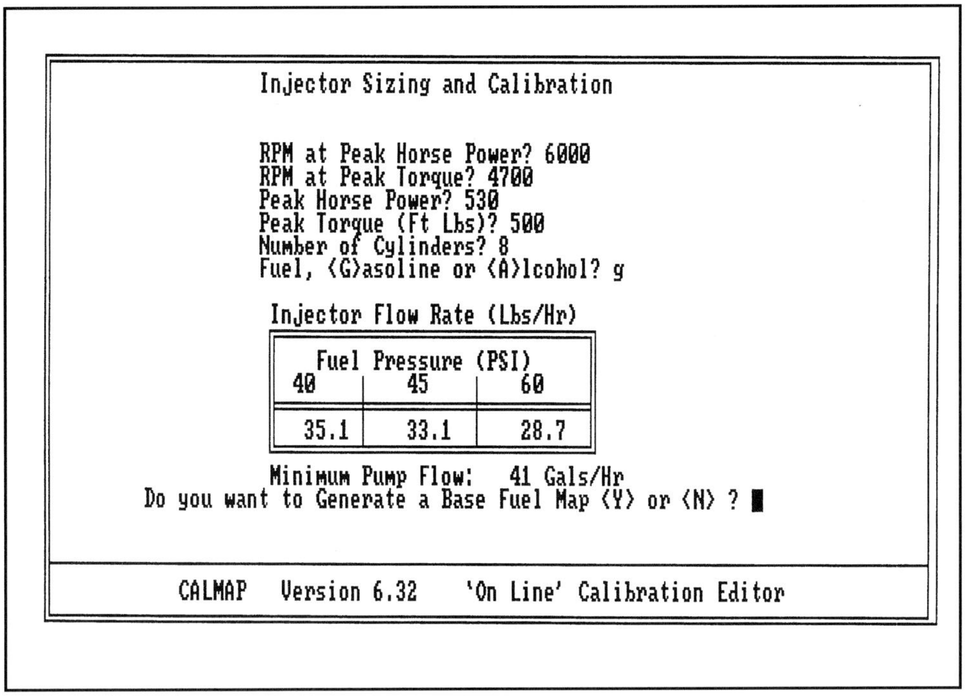

When doing a startup calibration, use the Utilities function in the Global menu to create a base fuel map.

enrichment is excessive. A very helpful screen when working to alleviate afterstart problems is the View function. The afterstart enrichment decay rate and percentages are easily viewed. To correct for a lean (stalling) condition, use a combination of increasing the initial afterstart enrichment at the appropriate coolant temperature; having the afterstart enrichment evoked sooner by reducing the number of engine revolutions in the beginning of the afterstart curve; or maintaining afterstart enrichment longer by increasing the number of engine revolutions in the decay rate curve. For the elimination of a rich afterstart program, lean out the amount of enrichment at the temperature where the problem exists; increase the number of engine revolutions in the beginning of the afterstart curve; or decay the enrichment out quicker by reducing the number of revolutions in the decay rate curve.

These same tables can be used to address hot-fuel handling problems that occur after a heat-soak of the engine. A start and stall after a heat-soak can be addressed best by enriching the afterstart curve from 180 degrees to 250 degrees F; minimizing the number of engine revs at the beginning of the afterstart enrichment curve; and delaying the decay rate. Another area that can affect afterstart stalling or rough running is the IAC starting position vs. coolant temperature. If the starting position is set too high, it is common for the idle to grossly overshoot on startup, and as the IAC position degrades, it can cause the engine to stall. The converse is that if there is too little IAC on startup, the engine will struggle to maintain rpm. Keep in mind that a fuel-injection system does not use a fast-idle cam to open the throttle plates to supply the additional air that is needed to compensate for the richer cold operation mixture. It relies strictly on the IAC.

Tuning Tips

Successfully tuning a fuel-injection system requires a broad-based background and an understanding of how all aspects of an engine are interrelated. Even though it would be impossible to anticipate every conceivable scenario that may be experienced, the following are the commonly asked questions to ACCEL's technical line. Very often a programming problem has its roots in more than one area and is the result of a stack-up of few different issues.

Idle Quality (Rolling or Surging)

One of the most common areas of difficulty is in obtaining good idle quality and a corresponding acceptable load-dump characteristic. Since most ACCEL ECUs are installed on modified engines, problems arise that would not normally be induced in a stock engine. The largest offender with regard to idle quality is the camshaft and intake manifold selection. As stated before, camshafts with excessive overlap do not interface well with long-runner intake manifolds and large plenum volumes. The runners and plenum act as storage for standing waves that are induced during reversion. What this causes is a tide-like effect of pressure variations in the plenum that are analogous to a liquid swishing around in a bucket, or fuel in a half-empty gas tank. These standing waves have a direct affect on cylinder fill and can also trigger MAP acceleration fueling. One of the easiest ways to stabilize the idle quality is to make sure the pressure variations are not evoking MAP fuel enrichment. Do this by accessing that menu and zeroing the first few points of enrichment. If that does not resolve the problem, try adding a small amount of idle spark to control the variation in engine speed.

Occasionally, a combination of the above with a slight increase in idle speed and a retarding of the base timing (through CalMap) will be all that is needed. It is also a good practice to use a vacuum hose that is six to 12 inches long to help isolate the MAP sensor from these pulses. Always keep in mind that injector repeatability at short pulse widths will impact idle quality and, if non-repeatable, will induce idle roll. Sizing an injector too large and trying to correct for it by decreasing pulse width will cause idle instability. Injectors that are issued pulse widths of less than 1.7 ms usually cause a rolling idle.

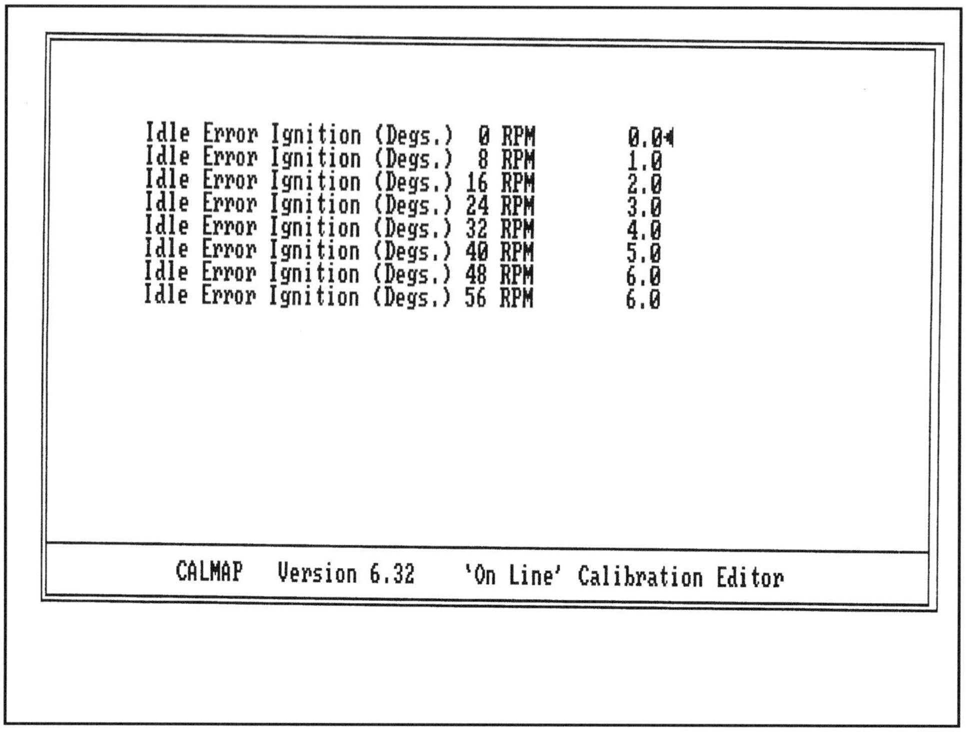

The idle spark function is evoked only above 100 degrees F engine coolant and below 1 percent TPS. The programmable number indicates the amount of spark advance for the level of idle variation.

Another common mistake is lowering the fuel pressure to accommodate a large injector. These attempts are usually successful as long as the pressure with vacuum on the regulator does not degrade below 35 psi, which is the threshold value for atomization. Even though pressure is always referenced with atmosphere on the regulator, we need to be concerned with the affect the engine vacuum has on lowering the fuel pressure. If the engine has very good vacuum (17–20 inches), then it will pull the fuel pressure down approximately 10 psi from the atmospheric reading. An atmospheric reading on the regulator of 40 psi could then become 30 psi on an engine with excellent vacuum.

Another tip for idle stability is to keep the cells around the idle point at the same or very near the same value as the idle cell. This will stop large changes in injector pulse width during interpolation. In some instances, if too much air temperature correction is programmed, idle instability will result. Once the idle is stabilized in Park, the issue of transitioning to a drive gear needs to be addressed. If the vehicle is using a stock torque converter, which is very tight, it will create a severe load-dump on the engine, wreaking havoc on your program. If an aftermarket camshaft is installed, the problem will become worse. Since the ACCEL ECU lacks the anticipation circuits of an OE design (most stock ECUs monitor gear position and command the IAC out to compensate for a load-dump from Park to a drive gear), it needs to see an idle speed degradation to evoke the Stall Saver function. A nice transition from Park to Drive or Reverse then becomes a delicate balance of fuel, idle speed, and Stall Saver functions.

Start by rocking the car from Park to Reverse and watching the fuel matrix to see the cell that it chooses. Often, richening the cell that the load evokes will cure a good portion of the stalling when placed in gear. Next, access the *Stall Saver* screen. Watch the cursor to identify how many counts of the IAC are being issued to compensate for that rate of idle degradation. The *Stall Saver* function is coded to hold the commanded amount of IAC steps for three seconds after acknowledgement of the engine speed drop. Raise the number of counts in the affected area to cure the problem.

Voltage Drop Due to Underdrive Pulleys—Another area that affects idle quality, and that even induces stalling and surging, is the lack of voltage that is available at idle with underdrive pulley ratios. This condition usually appears during an extended idle condition, such as at a long traffic light. It also is very electrical-load-sensitive, usually being induced at night with the headlights on or by a large power-hungry stereo. The only cure is either to idle up the engine, which is not the acceptable cure, or to switch to a pulley ratio that will maintain at least 12.2 volts during an extended idle with the electrical load on. A complaint about a stalling problem that is difficult to duplicate may be electrical load induced.

Using Fuel Pressure as a Tuning Aid

The use of fuel pressure as a tuning aid is very viable and needs to be addressed. The prescribed method for determining a fuel pressure starting point is to work from a 43.5-psi setting, which will deliver the rated capacity of the injector. Using the charts elsewhere in this chapter, the flow of the injector can be determined with a higher or lower setting. It is common to raise the fuel pressure to accommodate large fuel demands with a smaller injector. A maximum of 50 to 55 psi of pressure should be used with pressure rises above that, creating an increase in injector rise time and current draw on the ECU. Forced-induction engines need to maintain one psi of fuel pressure increase for each psi of boost to maintain the proper relationship of fuel pressure to manifold pressure. High-boost applications will place the fuel pressure over the above-mentioned threshold. Since most high-horsepower forced-induction engines usually require the use of a low-impedance injector, the available current to open the injector is greater, and the higher pressure has less of an affect on the rise time.

One issue that needs mention is the use of an FMU (fuel management unit) to dramatically raise fuel pressure under boost. Mostly the domain of supercharger manufacturers who are trying to sell their kits under the logic of not having to modify the fuel system, they are a very poor method to fuel an engine. Based upon a rate of gain, they skew injector rise times to such an extent that a high-rpm lean-out may occur. (Rate of gain is the pressure increase proportional to the amount of boost. It is common to have these units raise fuel pressure 10 psi for every pound of boost, yielding a total fuel pressure of 145 psi.)

The prescribed method, especially when an ACCEL ECU is being used, is to size the injector properly and maintain the boost/fuel pressure relationship. It is common to over-fuel an engine with the use of large injectors in both normally aspirated and forced-induction applications. The relationship of BSFC and injector sizing is often ignored, and excessive injector capacity installed. Due to the close proximity of the injector to the intake valve (usually 100mm in a OE manifold), it is very easy to wash the oil from a cylinder with too rich an air/fuel ratio. In comparison, this would be very hard to do with a carburetor. Wash-down causes not only the dilution of the engine oil, but rapid ring wear and, in turn, poor cylinder sealing. For this reason, an engine that is going to be calibrated should have the oil changed before the procedure begins, to prevent any residual hydrocarbons from affecting the programmed air/fuel ratio.

Tuning for Emissions

Trying to get an engine to pass an emissions test may require a slightly different technique than tuning with no regard to compliance. An engine equipped with a performance aftermarket cam will probably need to have less timing at idle to reduce hydrocarbon emissions. By calibrating as close to stoichiometric as possible, raising the idle speed, and maintaining closed-loop operation at low engine speeds, the engine should pass. A radical engine with a large cam and excessive crevice volume in the bores will be hard to get through HC testing if catalytic converter light-off cannot be achieved.

Throttle Dampening

The throttle follower function tracks the IAC position in regard to throttle angle and is used as a dash pot function to control stalling on rapid throttle closings. A common practice with enthusiasts is to free-rev their engines. If the throttle follower function is not correct, the idle speed will stay high when the throttle is released and will not respond as a carburetor would. To correct, this free-rev the engine and then release the throttle while accessing the *Throttle Follower* screen in the *Idle Speed* menu. Note the value the cursor travels to when the idle speed remains high, after release of the throttle. Lower the amount of bypass air at that position.

This function is also used on stickshift cars to eliminate stalling on deceleration while the clutch is still engaged.

Warm-up Enrichment and Air Temperature Correction

These functions can be used not only as trim characteristics to the base fuel map, but also as a tuning aid. If you are under the assumption that the engine wants to be richer in all areas of the fuel map and you just want to

Make Mine Zero

As shipped from ACCEL, the DFI system has values in all tables. Other than the timing and fuel maps, the calibration in the additional screens is generic and often not a good starting point for tuning. Areas such as the IAC functions, warm-up enrichment, acceleration fueling, cranking, and after-start fueling may not be correct for the use at hand. Most tuners end up using the supplied tables as a guideline and modifying the calibration from there. Even though at first this may appear to be a sound logic, it often is misleading since the required calibration may be completely different from the format supplied.

The best and quickest method to bring the ancillary calibration tables in line is to zero all the values and then build from there. As covered in the text, all of the tables except the warm-up enrichment and air temperature correction offer some coded input, even when the place values are set at zero. By doing this you are starting with a "clean sheet of paper" approach and are not limited by the implied boundaries of a factory-installed calibration.

try it out, access the *Warm-Up Enrichment* screen and add fuel as a percentage through this table. Conversely, if you would like to lean the whole program temporarily, then access the *Air Charge Correction* menu and lean the base fuel map out for test purposes by evoking enleanment.

Ignition Timing Requirements

ACCEL ECUs are equipped with a very conservative timing program as a starting point for your tuning. It is important to recognize that today's quick-burn combustion chambers do not require as much spark advance as the older cylinder head designs. Modern combustion chamber designs usually make the most power with a maximum of 36 degrees spark advance, instead of the customary 42 degrees of spark lead. Piston design also has a major affect on flame speed and, in turn, the amount of timing required. The key is to use your skills to optimize the timing curve and not input just one number for the complete matrix.

Acceleration Enrichment

This is used to supply fuel during very abrupt throttle changes that create a large drop in manifold vacuum. The TPS enrichment is considered an asynchronous pulse, which is an independent injector pulse that is inserted between the synchronous pulses. A synchronous pulse is defined as a continuous pulse repeated at the same rate. Access the *Acceleration Enrichment* menu and, while stabbing the throttle at different rates, follow the cursor and add or subtract fuel until the throttle response is crisp.

Working in concert with TPS fueling, MAP-based enrichment functions similar to a power valve in a carburetor. Once the Base Fuel Map is close to being correct, the MAP enrichment can be used to add crispness to the engine under load and as the throttle is crowded. If working the fuel matrix leaves undesirable results, such as the need to over-fuel a cell at a certain load input, utilize the enrichment function of the MAP sensor. This will help to deliver excellent tip-in load characteristics, eliminating any flat spots that may occur as the vehicle starts to climb a hill while cruising on the highway. For applications that are used for towing, this function is critical to prevent lean-out under extreme load while maintaining good fuel economy and emissions.

Idle Surge During Warm-up

If the engine experiences an idle surge during warm-up, it is most likely due to a lean air/fuel ratio. Use the Warm-up Enrichment Table to add fuel in the coolant temperature range where surging takes place. Keep in mind that the idle spark function does a very good job of controlling this idle roll but is coded not to be invoked until 100 degrees coolant temperature. Most of the idle stability problems during warm-up will be exhibited below the 100-degree range.

No-Start Troubleshooting

The first thing to remember when diagnosing a no-start condition is to check over basics. The ACCEL ECU is American-made, and that means it is extremely reliable and not prone to failure. Most no-starts after an installation are due to improper connections or incorrect hook-up of the ignition. Before condemning the fuel-injection system, always do a visual inspection, and don't forget to check the fuses for the ECU.

A good indicator that everything is working is if the ECU can communicate with your laptop. This will establish that there are at least 12 volts getting to the ECU. Access the *View* screen and scan all of the sensor data for proper inputs. A common error during installation is the wire from the main harness switched ignition source connected to a voltage supply that is available with the key on, but disappears during crank. When the ignition is turned on, you should hear the fuel pump prime for two seconds, and when going into crank you should be able to read rpm data on the View screen. Another possible problem area is the connection of the ECU ground someplace other than at the battery-negative terminal. By attaching this ground to another part of the vehicle, you are referencing a different ground plane, and in some instances this may cause a problem. It is important to keep in mind that on an EFI system, if there is no spark, there usually is no injector command.

The only time that this is not applicable is if the ignition circuit fails after the generation of the rpm reference signal (a faulty ignition coil winding). If the ECU does not receive a primary interrupt, then it is not aware that the engine is cranking and will not issue any injector commands.

If there is fuel but no spark, then the primary signal is being recognized by the ECU. The problem can most likely be found in the secondary side of the ignition.

A condition of spark with no fuel pulse needs to be addressed by accessing the View screen and checking if the ECU is recognizing the rpm signal. If the ECU is registering rpm, then check that the TPS function is not triggering clear flood mode and that the MAP sensor values are correct. Also, make sure to check both bulkhead connectors at the ECU for a tight fit and the individual wires exiting the connector for proper contact.

Dual Fuel Pumps

Due to the lack of access to some in-tank fuel pumps, it is common practice to leave the stock fuel pump in place and supplement its volume with an ACCEL fuel pump. Whenever using multiple fuel pumps, always make sure each pump is fed by its own voltage source with a relay. Electric fuel pump volume output ramps off dramatically with decreased supply voltage.

Nitrous Control

The last entry in the *Global* menu is the *Nitrous Settings.* This allows for programmable nitrous oxide control. The ECU has the ability to automatically add enrichment through the fuel injectors so separate plumbing is not required. It can also retard the ignition timing as a function of engine speed. This is very important because nitrous is an oxidizer and causes the air/fuel mixture to burn very quickly. The spark timing must be retarded to prevent detonation. The ECU also has the ability to select when the nitrous is initiated as a function of engine speed, throttle position, and delay time, once the criteria are satisfied.

Move the box down to select the NOS Settings and press the ENTER key. The screen will change to parameter settings. To change the values, use the arrow keys.

The only time you can make a change to a parameter is if the box is around the appropriate function. In other words, if you want to change the TPS threshold, the box must be around the Throttle setting.

The first box is the throttle position that must be exceeded to activate the nitrous. The range is from 50 to 100 percent. Type in the value you want and then press ENTER.

Now move the arrow key down to the next box. Once the speed and TPS threshold are exceeded, this is the delay time that the ECU will wait before activating the nitrous. The range is from 0.1 to 2 seconds. Type in the desired delay and press ENTER.

The third box is the speed threshold. This is the second threshold that must be satisfied (along with TPS) before the nitrous will be activated. The range is from 1,500 to 6,000 rpm. Type the desired setting and press ENTER.

The next is the percent of additional fuel that will be added through the fuel injectors. ACCEL chose this approach because the conventional method of adding more fuel resulted in wall wetting and distribution problems. By using the fuel injectors, the fuel preparation and distribution are very good. Nitrous boils at -129 degrees F when fed into the plenum and can be considered a gas, so the dry-flow characteristics of the intake manifold are maintained.

Stock Chevy 350 injectors are good for about 400 horsepower (total). The system will be pre-calibrated for a certain nitrous flow rate (metering holes in the plate). This function lets you increase the fuel flow through the nozzles. It should be understood that there is a limit to the fuel flow through the injector. That limit is reached when the injector is on continuously (static). At static flow, the only way to increase the fuel delivery is by raising the fuel pressure. If the decision is made to install a larger injector or raise the fuel pressure, you need to alter the entire Base Fuel Map to accommodate this change.

The fuel-delivery system must be

able to maintain at least 45 psi when the nitrous is activated. If this is not the case, the air/fuel ratio may become excessively lean, resulting in engine damage. If higher fuel pressure is used, check to see that it is maintained while the nitrous is activated. A fuel pump's ability to deliver decreases as the head pressure increases.

The final box is the spark timing retard curve. This allows retard of the ignition timing as a function of engine speed. The amount of spark timing retard is read on the y axis, engine speed on the x axis. It should be noted that while in this box, the right/left arrow keys allow movement on the graph. This curve is like any other in CalMap only it has less resolution. The break points are every 1500 rpm. However, there is linear interpolation done between set points. With this function you can modify the torque curve by varying cylinder pressure with spark timing. If the intake manifold makes less torque at low engine speeds (tunnel ram type), you would decrease the amount of spark retard (more advance) at low engine speeds. However, when the manifold comes into resonance, the spark timing will need to be retarded.

The ignition curve is subtracted from the Ignition matrix.

With the nitrous option, there is also an enable switch. Once enabled, this tells the ECU that the nitrous is anticipated. Once all the criteria are satisfied (rpm, TPS, delay time) the nitrous will be activated automatically while the spark timing is being retarded and the additional fuel added.

Knock, Knock

An engine equipped with a knock sensor is said to have closed-loop timing control. When detonation is detected, the ECU responds by retarding the ignition timing until the abnormal combustion ends. During this process, a phenomenon called "hysteresis of knock" takes places. The abnormal combustion event super-heats the combustion chamber and piston, which then requires an extreme amount of timing retard to cool down these parts so detonation does not reappear. For example, the ECU may need to remove 10 or 12 degrees of timing to cool the chamber from a calibration that is only one degree into detonation. The knock sensor cannot be thought of as an oxygen sensor, trimming the spark table as a fuel matrix. Any time detonation is detected, excessive timing retard will be issued, degrading power. The ideal when tuning a spark table is to never enter the hysteresis of knock, gently approaching the advance calibration for best performance without evoking abnormal combustion.

9

Tuning Tips For GM Engines

The majority of DFI product ends up on GM engines, especially those of the Chevrolet variety. The 6.0 ECU is a very common upgrade for both mild and heavily modified TPI and LT1 engines. The Buick GN and 1989 Turbo Trans Am are also very popular candidates, as are the GMC Syclone and Typhoon. The ZR-1 Corvette, with its unique ignition and variable-runner intake manifold, is not compatible with the 6.0 ECU. In addition, the popularity of Chevrolet engines in racing, marine, and street-rod applications makes the HEI-configured ECU ACCEL's most popular seller.

As mentioned throughout this book, the CalMap calibration software does not recognize the make or model of the engine it is running, but certain applications do require special procedures.

4L80E and 4L60E Transmissions

In these applications, the internal switching of the hydraulic fluids path that was normally accomplished by a mechanical valve body is replaced by select solenoids that are controlled by the factory engine-management ECU. For the transmission to shift, the ACCEL ECU needs to be piggybacked with the GM unit. The factory ECU looks at many engine parameters such as rpm, road speed, throttle angle, and coolant temperature to control the shift points, line pressure, and torque converter lockup. A jumper interface harness is offered by many EMICs that allows this integration to be a simple process. Another option is a programmable transmission controller. Units of this type were originally offered by ACCEL and worked independent of the 6.0 ECU. During changes in ownership of the Mr. Gasket Company, this product line was sold and either the same or a like product can be found marketed by ACCEL's competitors.

LT1

ACCEL offers an Opti-Spark-compatible ECU that works with all Generation II small-blocks. The LT1

Air Volume of Various Intake Component Ports
(All flow testing done @ 28 in. H_2O with approx. temp. @ 68°F.)

Runners (measured individually)

Stock	203.17 cfm
ACCEL	242.02 cfm
Extrude/ACCEL	275.83 cfm
Super Ram	289.18 cfm

Intake manifold w/3/8-in. radiused inlet
222.45 cfm

Stock intake manifold w/runner

Stock	198.72 cfm
ACCEL	213.52 cfm
Extrude/ACCEL	217.11 cfm
Super Ram	220.67 cfm

ACCEL Hi-Flow intake manifold w/3/8-in. radiused inlet 251.51 cfm

ACCEL Hi-Flow intake manifold w/runner

Stock	215.83 cfm
ACCEL	232.53 cfm
Extrude/ACCEL	243.21 cfm
Super Ram	240.24 cfm

Extrude Honed ACCEL Hi-Flow intake manifold w/3/8-in. radiused inlet 275.83 cfm

Extrude Honed ACCEL Hi-Flow intake manifold w/ACCEL runner 266.94 cfm

Edelbrock Performer RPM manifold (stock)
286.51 cfm

Edelbrock Victor Jr.
275.24 cfm

1989 Corvette head CFM
(stock head w/3/8-in. radiused inlet, 6-in. extension, $1^{3}/_{4}$-in.-dia. exhaust)

Intake cfm @ Inches of Lift
60.68 @ .100
117.28 @ .200
156.81 @ .300
176.47 @ .410 (stock 1989 Corvette cam lift)
177.97 @ .437 (1.6:1 rockers)
183.89 @ .499 (1.5:1 rockers, 211°/219° split duration ACCEL cam)
185.37 @ .533 (1.6:1 rockers, 211°/219° split duration ACCEL cam)
186.85 @ .550 (port stall point)

Exhaust cfm @ Inches of Lift
38.53 @ .100
99.84 @ .200
137.28 @ .300
156.32 @ .423 (stock 1989 Corvette cam lift)
157.90 @ .451 (1.6:1 rockers)
161.05 @ .525 (1.5:1 rockers, 211°/219° split duration ACCEL cam)
162.63 @ .560 (1.6:1 rockers, 211°/219° split duration ACCEL cam)

Intake/Exhaust Ratio @ Max. Lift

Stock	1.5	88.5%
	1.6	88.7%
211/219	1.5	87.5%
	1.6	87.0%

The Stealth BBC long-runner intake manifold is a popular conversion for large vehicles and often produces a drastic improvement in fuel mileage.

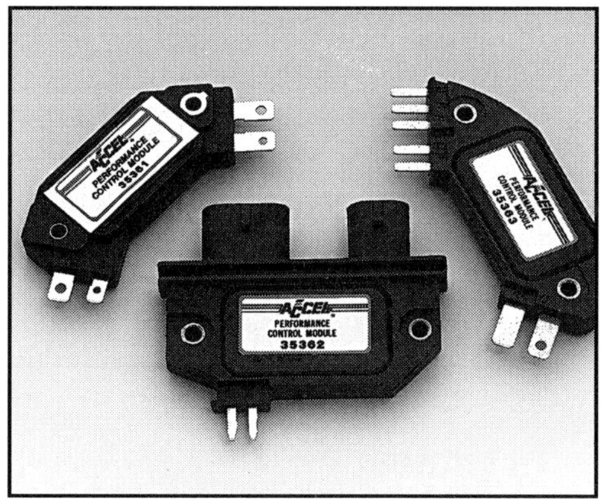

ACCEL offers a high-performance ignition module for almost every GM application.

A supercharged small-block like this one would use a two-bar MAP sensor for excellent driveability.

has no ability in factory form to make any mechanical timing adjustments other than through recalibration of the engine-management system. For this reason, the harmonic damper acts both as a vibration suppression unit and as the crankshaft pulley for the serpentine accessory drive belt. Due to this design, no timing marks or tabs are found on this engine. To synchronize the DFI with the actual coil firings, a mark and tab need to be fabricated on this engine locating TDC on the number one cylinder. This is best done while the engine is apart and with a dial indicator or

GM EFI Spotters Guide

There are more acronyms in use today in the automotive world than ever before in history, and General Motors electronic engine management has more than its fair share. Use this handy guide to make sense of this alphabet soup.

ALCL: Assembly line communication link
ALDL: Assembly line diagnostic link
BARO: Barometric absolute pressure sensor
CCC: Computer command control
C I: Computer-controlled coil
CCP: Controlled canister purge
CE: Check engine
CPI: Central port injection
CLCC: Closed-loop carburetor control
CTS: Coolant temperature sensor
DIS: Distributor-less ignition
EAC: Electric air control
EAS: Electronic air switching
ECM: Electronic control module
ECU: Engine calibration unit
EECS: Evaporative emissions control system
EFE: Early fuel evaporation
EMR: Electronic module retard
ESC: Electronic spark control
EST: Electronic spark timing
HEI: High-energy ignition
IAC: Idle air control
ILC: Idle load compensator
IP: Instrument panel
ISC: Idle speed control
MAP: Manifold absolute pressure
MAF: Mass airflow
MFI: Multipoint fuel injection
NC: Normally closed
NO: Normally open
PFI: Port fuel injection
PROM: Programmable read only memory
SFI: Sequential fuel injection
TBI: Throttle body injection
TCC: Torque converter clutch
THERMAC: Thermostat air cleaner
TPS: Throttle position sensor
TPI: Tuned port injection
TVS: Thermal vacuum switch
VSS: Vehicle speed sensor

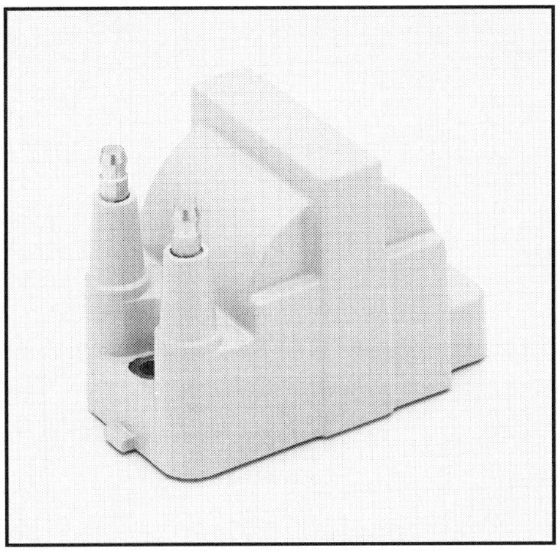

Normally aspirated engines with Type II DIS can enjoy the benefit of higher-voltage output from an ACCEL ignition coil, part number 140017.

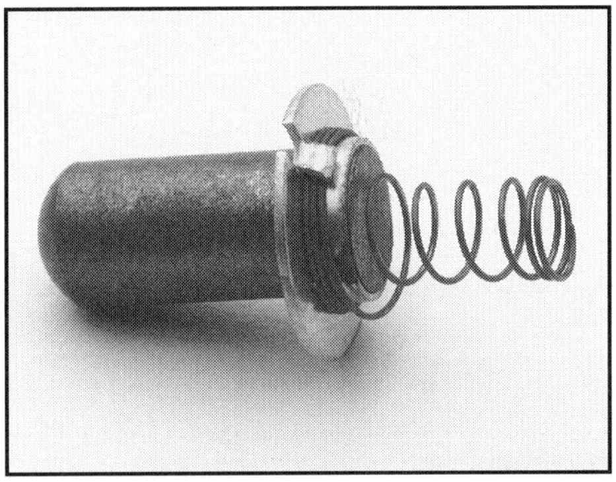

A high-performance coil button for HEI-equipped engines is a good choice for superior spark energy. This item is cataloged under part number 120900.

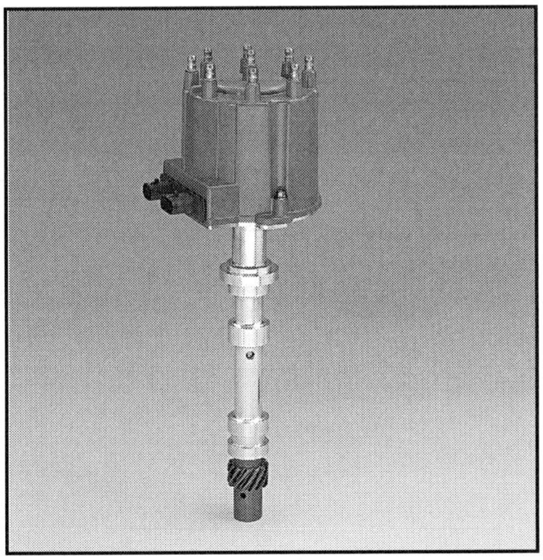

ACCEL offers a billet roller-bearing small-cap F-body-style distributor for use in any small-block Chevrolet. An option wiring harness makes this a drop-in for an older TPI engine.

positive piston stop and a degree wheel. If the engine is being converted to a DFI without being dismantled, a positive piston stop inserted through the spark plug hole can be used, but this usually creates a large amount of error.

The Opti-Spark cannot be moved to alter the static timing. Follow the procedure for synchronizing the timing (in Chapter 8), and instead of moving the distributor to have the value on the harmonic damper equal that in the box below the matrix, record the variation and use that to determine the actual spark timing of the engine.

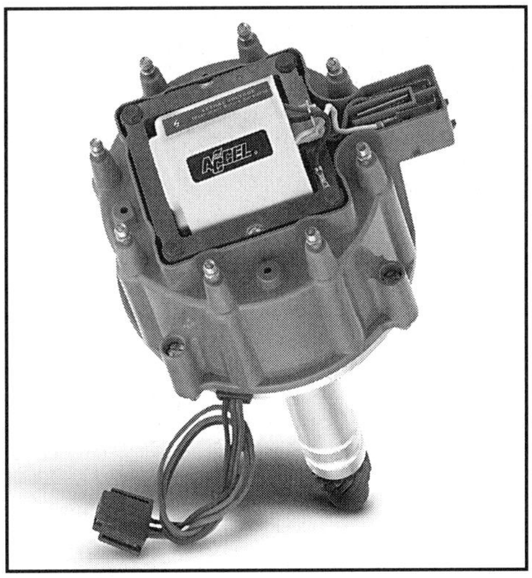

Don't forget that the ignition system is what starts the combustion event, and it needs to be up to the task before any tuning is done.

Buick Turbo V-6 engines use a Hall effect cam sensor to identify the cylinder events.

The Buick Turbo uses a Type I DIS coil. ACCEL offers a high-performance replacement under part number 140016.

A cold-start injector eliminator kit is offered under part number 74820.

Cold-air kits return the best performance for your investment dollar.

Ida Automotive tuned this carburetor look-alike DFI system in an older Corvette.

Buick Turbo V-6

The installation of a 6.0 spark/fuel ECU in a Turbo Buick engine converts the fuel injection from a sequential to a batch-fire system. Also, the mass airflow sensor (MAF) is eliminated and replaced with a MAP sensor. GM MAP sensors have a color-coded insert in the plug connector that identifies their scale. One-bar is black; two-bar is red; three-bar is yellow. If the car was originally equipped with an analog dashboard, then it has a two-bar sensor under the hood on the passenger-side fenderwell that operates the boost gauge. Digital-dash cars had a Hobbs-style switch that illuminated a low- and high-boost light in the instrument cluster. Buick controlled boost through the ECM with a duty-cycle solenoid. The 6.0-series ECU has no ability to control boost, so either a mechanical wastegate actuator or an external wastegate such as a delta gate must be used.

Timing Procedure

Due to the differences in this ignition system in relation to a conventional distributor, the timing verification procedure is unique. The ACCEL Spark/Fuel ECU for a GN is identified by part number 74025 and is compatible only with the Buick or ACCEL coil packs. The cam sensor must be set to the factory position of 25 degrees ATDC, and a Buick knock sensor and module need to be sourced from General Motors, but the ACCEL 74173 harness kit can be employed. Since there is no mechanical timing adjust on this engine, the procedure of generating a map of all the same values does not apply. For the Buick engine use the following steps:

1) Connect a dial-back timing light to the number one spark plug lead along with a PC or laptop with the CalMap software. Enter the CalMap ignition timing screen.

2) There will be a difference between the timing read on the harmonic balancer vs. the CalMap. The timing in the ignition matrix will be the timing command from the ECU, whereas the reading on the harmonic balancer with the timing light will be twice the actual timing plus an error of 12 degrees. For example, 52 degrees on the balancer would equate to a timing command of 52-12 = 40/2 = 20

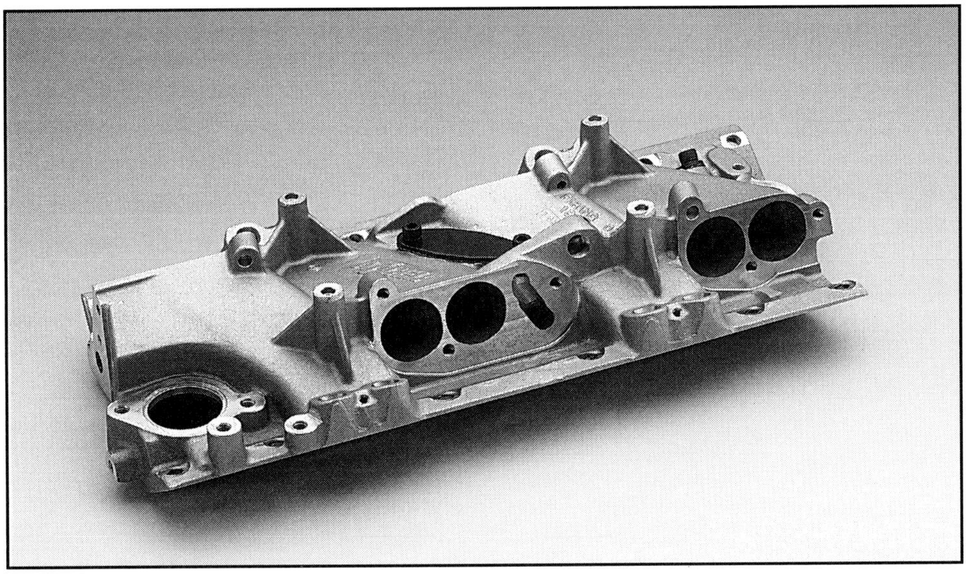

The Super Ram base manifold offers higher airflow than a stock TPI.

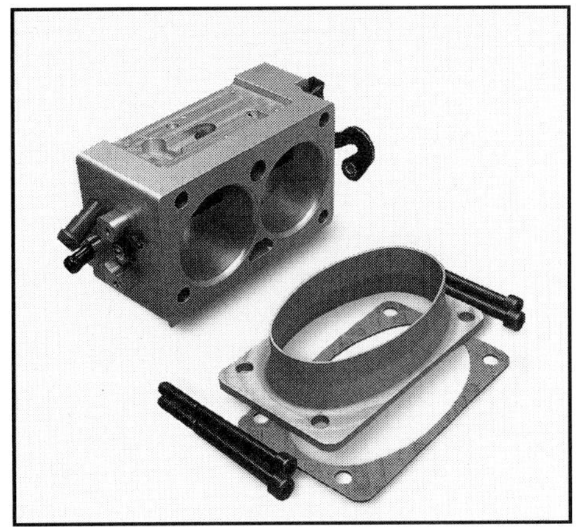

Made from billet aluminum, the ACCEL throttle body is a direct replacement for a stock TPI unit.

The GMC Syclone was hard to work on but made plenty of power. This one is getting upgraded to a Buick-style 3-bolt turbo by the crew at Dynotech.

The ACCEL single-plane intake manifold is an excellent choice for a high-rpm engine with a big camshaft and good cylinder heads.

degrees of spark advance.

3) The variation in the readings is due to the waste spark firing of this ignition system.

GN Coil Packs

Before tuning a Turbo Buick, always check the coil packs for resistance with an ohmmeter. It is common for the coils to fail when the engine is equipped with faulty ignition wires. The resistance should be below 13,000 ohms; a new coil will read approximately 11,700 ohms. Degraded coils will cause drivability problems such as bucking and backfiring under boost.

Syclone and Typhoon

These engines were produced by Chevrolet and were a modified version of the 4.3-liter V-6, which is basically a 350 with the rear two cylinders cut off. Due to this, the cylinder firing events are not the same as on a V-8, and the crankshaft crank pin angle is altered. The EFI system from the factory was a GM batch-style and adapts very nicely to DFI installation since it used a two-bar MAP sensor in its original form. As with the GN, the factory ECM controlled the boost level with a duty-cycle solenoid that is not a part of a DFI.

Buick Turbo Distributorless Ignition Systems

As good as GM High Energy Ignition was, it still used one coil to fire all the cylinders and was triggered by a distributor that encompassed numerous moving parts. Since spark timing is referenced from crankshaft position, it never made any sense to identify this through the intermediate components of a timing chain and distributor gear. This introduced added tolerance and error into the spark advance curve, and is the reason behind race engines using crank triggers. GM DIS identifies the crankshaft location with a Hall-effect switch triggered by the harmonic balancer. DIS incorporates multiple coils, each of which has the responsibility to fire two cylinders, leaving more time for charging and cool-down. The additional coil saturation time, along with the integral electronics, allows for higher current and voltage output over conventional ignition systems. In addition, on average, a GM DIS can charge the ignition coil with 8.5 to 10 amps of current to an HEI's five amps and a points ignition's three amps.

The most unique aspect of the Buick C^3I system is its use of a waste spark. In this design, each end of the coil secondary is attached to a companion spark plug. This is defined as having a piston at TDC simultaneously, one on compression and the other on exhaust. The bore on the compression stroke is identified as the event, and the exhaust cylinder is considered the waste. When the coil discharges, both spark plugs fire at the same time, completing a series circuit. Since the polarity of the primary and secondary windings is fixed, one plug will always fire from the center electrode to ground, whereas the companion will do the opposite. This differs from a conventional ignition system that always fires the spark plugs from center to ground electrode. A Buick V-6 has a firing order of 1-6-5-4-3-2. This system requires that cylinder number 1 be identified and, seeing this synchronization signal, start the engine on companion cylinders 6 and 3. When the coil is discharged, the polarity is determined by the cylinder pressure. Since high cylinder pressure requires more spark energy, this will become the dominant event and fire the plug with a conventional electrical path.

After reading this, an inquisitive mind may be thinking: If one secondary lead is disconnected, then it must kill two cylinders. It may seem that way, but that's not how it works. Since the two plugs are in series with the coil secondary windings, both fire at the same time. The waste firing requires little voltage, leaving sufficient electrical power to fire the event plug. It’s possible for one plug to fire with the companion secondary lead disconnected. What happens is that the disconnected spark plug lead acts as one plate in a capacitor, with the engine itself being the other plate. These two capacitor plates are charged as a current surge jumps across the gap of the connected spark plug. The plates are then discharged as the secondary energy is dissipated in an oscillating current across the spark plug gap.

DFI in Wixom, Michigan, does calibration work for Ford Industrial Engines. Here, a Triton V-10 undergoes evaluation.

10

Tuning Tips For Ford Engines

The most common Ford you'll find equipped with a DFI system is a race-dedicated 5.0-liter Mustang. The early EFI 5.0 engines were equipped with a speed-density-based EEC-IV (electronic engine control, fourth generation) system. During the late 1980s, Ford switched to a mass airflow (MAF) design that gained popularity very quickly among enthusiasts of this make. Ford also offered, through its Motorsports division, a kit to convert speed-density engines to MAF. Quickly the aftermarket learned that alterations to the calibration of the MAF allowed injector size changes with no internal rework of the EEC-IV tables. This was in contrast to the GM design that required a chip or PROM change if the injector size was altered. In addition, the Ford electronics were much more advanced than those on the GM system of the same era, making it very adaptable to cylinder head, camshaft, and other internal engine modifications. Due to this, the ACCEL system did not gain the popularity among the Blue Oval boys as it did with the GM crowd. During this time, Ford SVO released a tuner program and piggyback setup that offered its own calibration software, which allowed the retention of the Motorcraft electronics and end user reprogramming. Thinking like this, and Ford's very proactive stance in the performance marketplace, has allowed it to trounce its competition and force General Motors to discontinue the Camaro and Firebird.

The weak link in the Ford ECU is the driver's inability to run low-impedance injectors for any length of time, forcing the use of an aftermarket EFI controller, such as the ACCEL DFI, on high-horsepower race entries. Many drag-race Mustangs use the

The 2.0L Zetec engine used in the Escort ZX2 and Focus has an EDIS distributor-less ignition and is not compatible with the 6.0 ECU.

ACCEL makes both adjustable and nonadjustable fuel pressure regulators for 5.0-liter Fords.

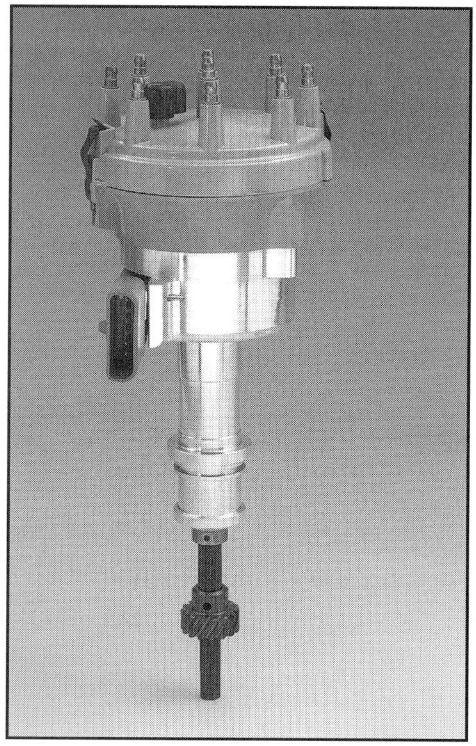

An ACCEL Billetech roller-bearing TFI distributor is a direct drop-in for DFI or EEV-IV systems.

Most Mustangs that run EFI and nitrous are configured in alpha-numeric.

ACCEL system, and its popularity has been especially strong with the nitrous crowd. The 6.0's ability to totally integrate all aspects of nitrous control has made it a must-have in the ranks of Pro 5.0 racing.

Installing a DFI System

There are some unique aspects to installing a DFI on a Ford engine, especially one that was originally EFI, and the following identifies these areas.

MAP Sensor

Speed-density-equipped EEC-IV systems used a frequency or square-wave-output MAP sensor. This is in contrast to a GM sensor, which is analog in output. Due to this

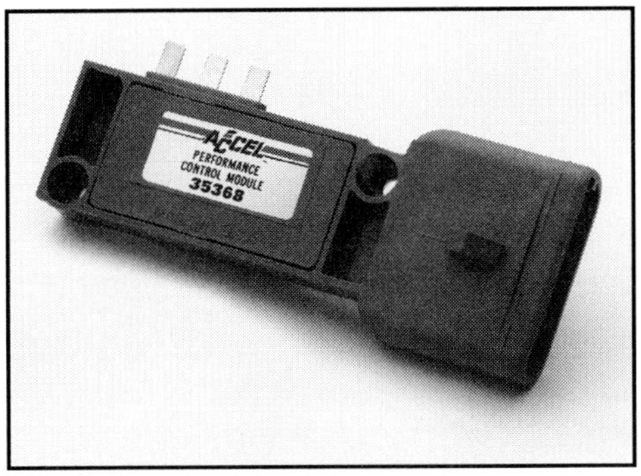

A high-performance replacement TFI module is offered under part number 35368.

EDIS ignition systems can be upgraded to a high-performance ACCEL coil with part number 140018.

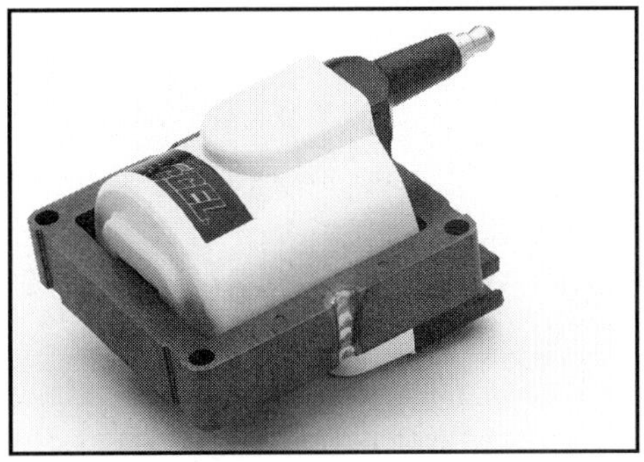

TFI applications are covered with part number 140012.

difference, any Ford installation requires the use of a GM/DFI MAP sensor, since the ECU is not able to read a frequency signal on this circuit.

Coolant Sensor

Once again, the Ford-issued sensor is not compatible with the GM-coded DFI ECU. The ACCEL-supplied sensor needs to be installed in the stock location.

Timing

Follow the set timing procedure outlined in Chapter 8, but instead of using a map of six degrees, Fords require a calibration of 10 degrees BTDC. If a six-degree test map is used, the timing will always be four degrees off from the value inserted in CalMap.

IAC

The ACCEL system uses an IAC motor that is GM-derived and is a stepper design. Fords use a dithering duty-cycle valve that works in a similar fashion to a Bosch idle-speed controller. The DFI system has no ability to duty-cycle this controller, so an adapter is offered to attach the stepper motor IAC to a Ford throttle body.

TPS

Early 5.0-liter engines used a TPS with a small slot, allowing a minute degree of adjustment; later models had no tuning features. ACCEL offers a modified Ford TPS that has the screw holes elongated and made into a slot for adjustability. This is the same throttle position sensor ACCEL used on its Harley-Davidson conversion kit and is often referred to as a Harley TPS. A stock Ford TPS can be modified with a file or die grinder to allow for the adjustability required with CalMap.

When a 6.0 DFI is installed, the Ford mass airflow sensor is removed from the induction track.

The EDIS ignition system fires a waste spark and works in a similar way to the DIS found on a Buick Turbo V-6.

Ford uses a duty-cycle or dithering idle speed controller. ACCEL offers an adapter to convert to a GM stepper motor IAC.

The Ford TPS can be used but may need slots cut into the mounting holes for adjustment.

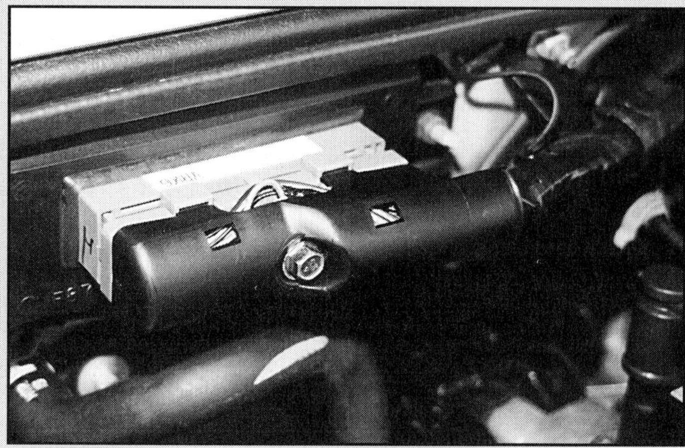

A jumper harness can be made to utilize the Ford wiring harness with an ACCEL/DFI.

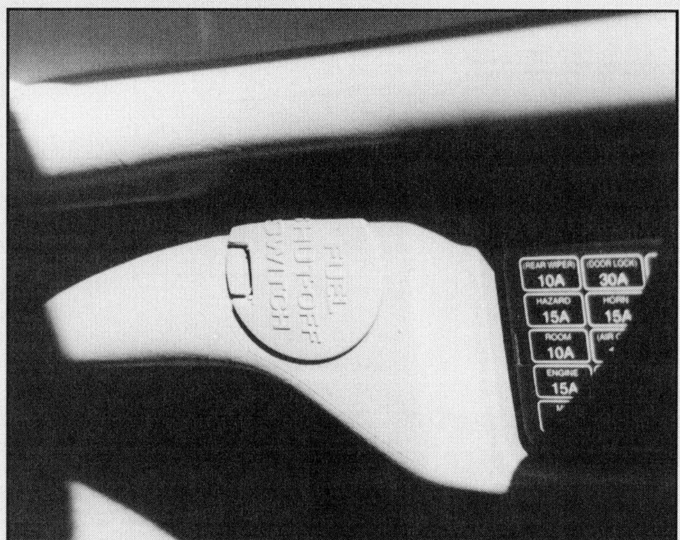

All fuel-injected Fords have an inertia switch that opens the power supply to the electric fuel pump if the vehicle gets bumped in an accident.

A highly modified Honda will require the tuneability of a DFI system for optimal performance.

11

Tuning Tips For Honda Engines

The import performance market is driven primarily by Honda, and specifically by the VTEC engines. The installation of an IPU-configured 6.0 ECU has become a popular choice for the enthusiast who has seriously modified his engine with either forced induction or nitrous. Honda's antiquated thinking in regard to ignition is represented by its use of a distributor. This has allowed the integration of a DFI system with relative ease. Most new-design engines use some style of distributorless ignition, and would then need to be ignition-compatible with the ACCEL ECU.

ACCEL does not build a four-cylinder-specific wiring harness, so an eight-cylinder version needs to be used and modified for smaller engines. This can be accomplished by removing the covering and snipping back and shrink-wrapping four of the injector's leads. Keep in mind that each ECU driver on an eight-cylinder engine operates two injectors, and the wiring harness is wired to the ECU in two banks. By studying the 015013 pin-out list in this chapter, the injector wiring harness can be easily modified so that each driver on a four-cylinder application will operate one fuel injector. This is extremely important since a highly modified forced-induction engine has a very elastic fuel demand. During normally aspirated operation, the injector pulse width will need to be tailored to the minuscule fuel requirements, but under boost large quantities of gasoline are needed. By altering the injector wiring harness in this manner, high-flow low-impedance fuel injectors can be used, and they maintain a clean idle and low-speed operation without the integration of a VIC, in most cases.

Most Honda applications for street usually require having the ACCEL ECU piggyback with the stock engine-control system. This is because the charging circuit is controlled by the OE computer, along with an incompatible idle-speed controller. The 6.0 ECU can then be wired to control ignition timing and fuel delivery, while the Honda electronics handle ancillary functions.

For the IPU circuit to acknowledge

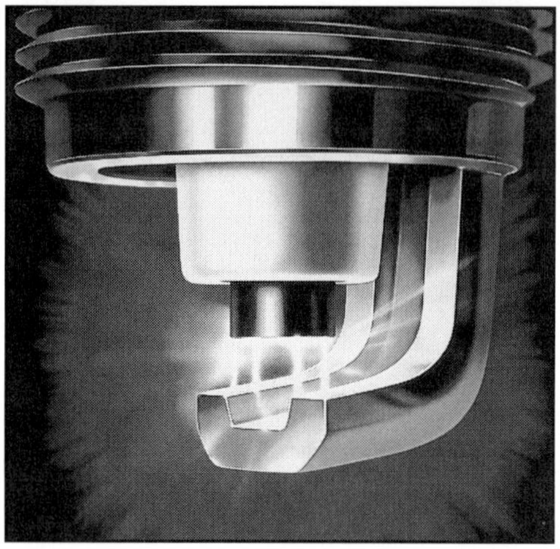

Boost means nothing without an efficient ignition event. The factory-issued Honda spark plugs usually don't work well with forced induction. Replace them with a high-quality ACCEL spark plug.

Part number 49310 is a plug-in conversion kit for the arc-welder-like ACCEL 300+ ignition system.

a sine-wave signal from either a distributor or a crankshaft sensor, the output needs to be a minimum of 1.6 volts AC peak to peak. This describes the amplitude of the sine-wave output from the negative to the positive side of the zero crossing. In my experience, a sufficient signal is available from most, but not all, Honda distributors. It may also be necessary to drill a small hole in the distributor cap to gain access to the wire from the signal generator for the ACCEL ECU. If this is done, make sure a grommet and silicone sealer is used to keep moisture from entering the distributor.

The Honda coolant sensor has a different calibration and threads from the GM-supplied part. The Honda sensor can be retained, but the wiring harness from the DFI will need to be altered since the plug is different. Simply remove the DFI Packard connector and solder in the Honda wires. The coolant sensor will read with a small amount of error, but as long as this is acknowledged, it can easily be compensated for in the appropriate CalMap screens.

The Honda TPS can be integrated

The 200+ is a good entry-level ignition for a mildly tuned engine.

into the ACCEL ECU in a like manner. If this is done, make sure that the polarity is correct. To check, enter the *View* screen after the conversion and watch the throttle position reading. Reversed polarity will have the screen read 99.9 percent at closed throttle and go down to zero at WOT. If the Honda sensor cannot be adjusted to read 0.5 percent at idle, do not be concerned. Any reading over one percent will only negate the idle spark function, and since a small-displacement engine will have a high idle speed, this control is not required.

If the Honda ECU is not being retained, and the ACCEL system will be the only management system, then idle-speed control is not available. Once again, this is not an obstacle since all that is required to compensate for the loss of the IAC function is to raise the idle speed to approximately 1,100 to 1,200 rpm.

VTEC-equipped engines can retain this function and actually tune it using the DFI system in a unique way. By wiring the power circuit to the B9 terminal of the DFI, as would be done in a nitrous installation, the rpm, throttle position, and amount of delay before the valve timing and lift change can be tuned.

Before any work is done, make sure you have the proper wiring diagrams for your year and model engine and car.

The following terms have been chosen for their relevance to the subject matter of this book.

A

advanced ignition: The timing event in a spark ignition engine that occurs before the optimal setting.

air-bleed passage: Jet or passage in a carburetor with a branch to atmosphere through which air is drawn and introduced into the fuel flow.

air-check valve: One-way check valve in an emissions-control air-pump system that prevents return of exhaust gases to the intake system.

air/fuel delivery ratio: The mass ratio of air to fuel inducted by an engine.

air injection: The addition of an air stream under pressure, but particularly to an exhaust system to promote combustion of unburned hydrocarbons and the conversion of carbon monoxide to carbon dioxide.

air silencer: Device placed at the entry to an induction system to attenuate the noise of induction.

anaerobic: Without air, in the sense of a process that takes place with the exclusion of air.

anti-backfire valve: Valve that allows air to flow from an air pump into an induction system on deceleration to prevent backfiring.

antiknock: Counteracting a propensity for abnormal combustion in an engine. Can refer to fuel or combustion-chamber design.

antiknock index: Specifications of the antiknock properties of gasoline in North America. Defined as one-half the sum of the research and motor octane numbers of gasoline.

aromatics: Unsaturated hydrocarbon additives containing at least one benzene ring, such as toluene, xylene, and elated hydrocarbons, which may be added, as appropriate, to fuels or lubricants.

ash: Non-combustible residue, or a fuel after combustion.

atomization: Conversion of a liquid into very fine particles.

automatic choke: Thermostatically operated choke valve in a carburetor inlet track that closes when cold and opens gradually as the engine temperature rises.

B

backpressure: Pressure resisting the flow of gas in an exhaust pipe.

backfire: An explosion of unburned or partially burned fuel in an exhaust or inlet system.

ballast resistor: Electrical resistor used to regulate ignition coil output at higher engine speeds and increase spark voltage during starting.

base blend: Primary liquid contents of a fuel to which additives are added.

base stock: Primary liquid constitute of a lubricant to which additives are added.

benzene: Organic chemical consisting of a hexagonal ring of carbon atoms with a hydrogen atom attached to each. Compounds containing one or more benzene rings are called aromatics and are an important part of petroleum.

benzole: Mainly aromatic hydrocarbon fuel additive or solvent.

blowback: Sudden reversal of airflow through a carburetor, often caused by incorrect ignition or valve timing.

blow-by: Unwanted leakage past piston rings of an engine.

blow-through: A forced-induction system for an engine that has the compressor blow directly into the carburetor, thus having the carburetor operate under a positive pressure.

boost: The additional pressure above atmospheric that a blower creates.

bore/stroke ratio: Ratio of the engine's bore to the stroke. A ratio of 1:1 is considered a square engine.

bottom dead center: The point of piston travel when the piston is nearest to the axis of the crankshaft. The lowest point of travel for the piston in the bore.

brake mean effective pressure: Engine cylinder pressure, derived by calculation, that would give the measured brake horsepower. It is the product of indicated mean effective pressure (IMEP) and mechanical efficiency.

brake power: Power developed by an engine as measured at the crankshaft by a dynamometer brake.

brass: Alloy of copper and zinc, often with other elements and proportions of zinc up to 40 percent. Grades exist with good electrical and thermal conductivity and moderate strength.

breather: A vent to an enclosed container or case, for example, a fuel tank or engine crankcase.

breeches pipe: Y-configuration exhaust pipe forming a connection of two exhaust manifolds to one exhaust pipe.

butterfly valve: Disk valve pivoted about its diameter and acting as a throttle in a pipe or chamber. Used as a carburetor term to describe throttle plates but is a misnomer. An actual butterfly valve consists of two half-hinged disks, opening and closing like the wings of a butterfly.

bypass valve: A valve for redirecting flow through a bypass.

C

cam follower: The part of a cam mechanism that rides on the contour surface of a cam.

cam profile: The shape of the periphery of a cam. The contour determines the stroke and linear action of the follower.

cetane: Paraffinic hydrocarbon and primary reference fuel to determine the ignition quality of diesel fuel.

CFR engine: Variable-compression test engine developed to determine the antiknock properties of gasoline.

charge air cooling: Cooling of the air charge downstream of a blower.

charcoal canister: Trap containing charcoal granules to store fuel vapors and prevent them from entering the atmosphere.

charge: The quantity or mass of the fuel and air entering the cylinder of an engine.

charge cooling: Removal of heat from the induction track of an engine to increase the density of the charge.

choke: Valve that restricts the amount of air entering an engine on the induction stroke, thereby enriching the air/fuel ratio for ease of starting and running when cold.

choke stove: Heat-exchange chamber on the exhaust manifold to speed the operation of an automatic choke.

choke valve: Valve that restricts the flow of a gas or liquid. Sometimes referred to as a choke plate.

choked: An engine running with the choke in operation.

CI engine: A compression-ignition engine commonly known as a diesel.

closed-loop engine-control system: Automatic control of engine parameters through direct feedback of data, as from electronic sensors.

closed-loop fuel system: Electronically controlled carburetor or fuel injection system in which the air/fuel ratio is adjusted by a feedback signal from the monitoring of exhaust gas composition.

coke: Carbon deposits in engine.

cold cranking rating: Current drain of battery when the starter motor is turning a cold engine for a specific time.

cold plug: Spark plug that operates at a low temperature in relation to the combustion temperature, having a relatively high heat conductivity.

cold-start device: Any device, such as a choke or additional fuel injector, that temporarily enriches the flow of fuel to an engine to facilitate starting when cold.

combined fuel economy: Standard measure of fuel economy derived from a stipulated proportion of urban and highway cycle driving.

combustion chamber: The part of an engine in which combustion takes place, normally the volume of the cylinder between the piston crown and the cylinder head casting.

combustion chamber surface-to-volume ratio: Ratio of surface area of the combustion chamber to its volume at a stated stroke of the piston. The ratio influences the quenching effect of the cylinder walls and the local formation of hydrocarbon emissions.

common rail: A fuel-injection system that has the fuel supplied to each injector from one high-pressure source.

compensating jet: Jet or passage in a carburetor with a branch to atmosphere through which air is drawn and introduced into the fuel at high flow rates, thus preventing an overly rich mixture.

compound carburetor: Carburetor with more than one choke or mixing chamber per inlet port.

compression: The increase of pressure in an engine cylinder as the piston travels toward top dead center.

compression-ignition engine: Reciprocating engine in which the fuel charge is ignited spontaneously by the heat of combustion. A diesel engine.

compression ratio: The ratio of cylinder combustion chamber volume at the bottom of the stroke to the value at the top of the stroke. Compression ratio is a volumetric ratio and not a pressure ratio.

compression ring: The uppermost piston ring.

compression stroke: The stroke of a reciprocating engine in which the air or air/fuel charge is compressed prior to ignition.

crankcase: Part of the structure of an engine that contains and supports the crankshaft and main bearings.

crankcase compression: Induction process in some smaller two-stroke engines, when the mixture charge is compressed in a sealed crankcase by the descending piston prior to passing the combustion chamber by way of a transfer port.

cranking enrichment: Additional fuel for starting a fuel-injected engine.

cross flow: A cylinder head having the intake and exhaust manifolds on opposite sides, creating a flow path across the cylinder head.

D

deceleration valve: Valve that allows extra air to flow into an intake manifold on deceleration to prevent backfiring.

delivered air/fuel ratio: The ratio of the mass of air to the mass of fuel, delivered to, but not necessarily combusted within, the cylinder.

detonation: Rapid and uncontrolled combustion. Detonation can occur in the cylinder of a spark ignition engine when operating on a fuel of inadequate octane rating, or with ignition timing too far advanced. Also referred to as pinging, knock, and abnormal combustion.

diaphragm carburetor: Float-less carburetor incorporating one or two diaphragms and capable of operating at any angle and normally unaffected by vibration and acceleration.

diesel engine: A reciprocating engine that operates on compression ignition.

dilution air: Used in exhaust emissions testing, ambient air that is passed through filters to stabilize the hydrocarbon concentration and dilute the vehicles emissions.

DIN rating: Standard for measurement of engine performance specified by the Deutsche Institut fur Normung (DIN).

direct injection: Diesel engine injection system in which the fuel is injected directly into the engine cylinder and not a pre-combustion chamber.

director plate: Multiple orifice plate fitted to a gasoline fuel injector to aid in spray formation.

discharge tube: Tube by which an emulsified air/fuel mixture enters the venturi of a fixed-choke carburetor.

double-barrel carburetor: Carburetor in which two barrels share the same float chamber.

downdraft: A carburetor through which the intake air flows vertically downward into the manifold.

Dual-fuel engine: An engine capable of running on two distinct types of fuel.

E

economizer: Device that regulates the fuel flow of a carburetor under high load conditions. More commonly referred to as a power valve.

emulsion: Mixture of fine droplets of one fluid dispersed in another fluid. Also, partially vaporized and heavily enriched air/fuel mixture prior to entering the main venturi of a carburetor.

emulsion tube: Combined main and compensating jet tubes in a carburetor with provision for drawing air into the fuel flow to create an emulsion at higher engine speeds.

engine map: Three-dimensional graphic representation of engine control parameters, for example, angle of ignition advance plotted against functions of throttle opening and engine speeds.

enrichment: An increase in the amount of fuel to a fixed amount of air in a carburetor.

ethanol: Ethyl alcohol as an alternative fuel or as a composite of gasoline.

evaporative emissions: Fuel vapors that escape to atmosphere by evaporation.

excess air factor: Factor by which the air/fuel ratio of an inducted mixture exceeds that of stoichiometric (excessively lean).

exhaust analysis: Quantitative measurement and presentation of the constituents of an engine's exhaust gases, vapors, and particulates.

exhaust backpressure: Resistance that impedes the flow of the exhaust gases from the engine to atmosphere.

exhaust calorimeter: Device for measuring the heat content of an engine's exhaust.

Exhaust-gas recirculation: Mixing of exhaust gas with intake air to decrease the potential specific heat of the charge and the formation of NOx.

exhaust stroke: Motion of the piston in an internal-combustion engine that expels spent gases from the cylinder. Also identified as the pumping loop.

F

fast idle: The high-speed idle used on a choked engine.

firing order: The numbered sequence in which the cylinders of a multi-cylinder engine fire.

fixed-choke carburetor: A carburetor with a constant-size venturi, with airflow controlled by a throttle plate.

flash point: The lowest temperature at which the vapors of a flammable liquid product will ignite, though not necessarily remain under continuous combustion.

flat spot: A transient reduction in torque of an engine on acceleration, often called a hesitation.

float: A buoyant part of a fluid-metering system.

float bowl: Main fuel reservoir in a carburetor.

float needle: Needle valve actuated by a float in a carburetor.

flooding: Condition that prevents starting of an engine when more fuel is drawn in than can be ignited.

four-stroke cycle: Thermodynamic cycle of an engine that requires four strokes of the piston to be completed.

friction drag: Aerodynamic drag resulting from friction between the moving air and the surface of the vehicle.

front-end octane: Research octane quality of the lower-boiling-point components of gasoline.

fuel: A combustible form of energy for an engine, usually in liquid form.

fuel cell: Device that converts gasoline or liquid fuel into electrical energy through electrochemical rather than mechanical forces.

fuel consumption: Rate of consumption of fuel by an engine expressed in units, such as miles per gallon.

fuel injector: Device whereby fuel is injected in metered quantities into an engine.

fuel-pressure regulator: Pressure-actuated diaphragm valve that maintains the pressure in a fuel system to a preset value above manifold pressure.

fuel pump: Mechanical or electrical device that draws fuel from a tank to provide the fuel supply to the carburetor or fuel injection system.

fuel-flow filter: Filter through which the total flow of fluid passes, as opposed to the bypass filter, which filters only part of the flow.

G

gas chromatography: System of analysis of gases or vaporized liquids, such as exhaust gas, using the chromatograph, which separates and identifies individual components of a mixture according to their tendency to be absorbed in a column or a capillary tube.

Gasohol: Automotive fuel normally consisting of nine parts gasoline to one part alcohol.

gasoline: Light hydrocarbon fuel used in spark-ignition engines.

gauze carburetor: Early form of carburetor in which induction air gathered fuel vapor from gauze over which gasoline flowed. Also known as a wick carburetor.

gross power: The measured power output of an engine operating without any power-absorbing accessories such as an alternator, generator, pumps, or induction silencers.

gulp valve: Valve to introduce extra air to induction track on acceleration to prevent an overly rich air/fuel ratio.

H

harmonic-induction engine: Induction system in which the length of the inlet track is chosen to improve volumetric efficiency over a narrow operating speed.

heat-control valve: Valve that regulates the flow of exhaust gas so that some of its heat content is passed to the intake manifold, thereby helping to vaporize the fuel mixture in a cold engine.

heat engine: Engine deriving energy from the heat of combustion of a fuel, whether burned internally of externally.

heat range: Range of temperature for optimum operation of a spark plug.

heat range index: Standard for the designation of the range of operating temperature of a spark plug.

heated intake: System whereby the induced air or air/fuel mixture in an engine is heated to reduce emissions on starting to allow operation in the cold.

horsepower: The customary non-metric unit of power, equivalent to 0.7457 kilowatt, defined by a rate of working of 33,000 lbs./ft. per minute.

hot-soak losses: Fuel vapors emitted during a specific period beginning immediately after the engine is turned off.

hot-start enrichment: Enrichment of the fuel mixture for starting a hot engine, usually applicable only to gasoline engines.

hunting: Variation of speed of an engine at a constant fuel setting.

hybrid engine: Engine combining two principal modes of operation, such as that of a diesel engine and spark-ignition engine.

hydrocarbon emissions: Unburned hydrocarbon-based fuel emissions from an internal-combustion engine.

I

idle speed: Rotational speed of an engine during a no-load or minimal throttle setting.

idle system: Arrangement of jets and tubes in a carburetor to enrich the fuel supply when the engine is idling.

idling jet: Slow-running jet.

ignition: Initiation of combustion.

ignition coil: Induction coil or voltage transformer that provides the high-tension voltage for the spark in a spark-ignition engine.

ignition delay: Time interval between the spark and the initial release of heat of combustion in a spark-ignition engine.

ignition timing: The timing of the spark relative to the piston top dead center in a spark-ignition engine.

IHP: Indicated horsepower

indicated power: Engine power calculated from an indicator diagram. Power developed in the cylinders rather than the output shaft of the engine.

indicated thermal efficiency: Ratio of indicated work available at the piston to the ideal work available from combustion. The thermal efficiency assuming zero friction and pumping losses.

induction: Drawing air or an air/fuel charge into an engine.

induction air: Air drawn into an engine before the introduction of fuel.

induction stroke: The stroke of the piston in which the working fluid is drawn into the cylinder.

inductive kick: Phenomenon that augments the secondary voltage in an ignition system by interrupting the current after the points have been opened.

instantaneous piston velocity: Piston speed at any specific point in its stroke or crankshaft angle.

intake depression: Mean reduction in static pressure below atmospheric in an engine air-intake system.

iso-octane: Hydrocarbon used as a primary reference fuel in determining the octane of fuels, having assigned values of research and motor octane.

J

jackshaft: Small shaft within a machine for transmitting rotary motion, for example, in an ignition distributor.

jet: An accurately drilled hole through which liquid can pass at a controlled rate, as in a carburetor.

judder: A low-frequency vibration.

K

Kettering ignition system: Commonly used inductive-ignition system comprising an ignition coil, breaker, capacitor, and battery pioneered by Charles Kettering.

knock: Noise resulting from the spontaneous ignition of a portion of the air/fuel mixture in an engine cylinder and occurring ahead of the normal spark-initiated advancing flame front.

knock rating: Octane rating

knock sensor: Instrument that detects the onset of detonation in an internal-combustion engine.

L

lambda sensor: Electrochemical sensor that relays data on oxygen content of exhaust gases to an electronic engine-management system or laboratory equipment.

lean-burn engine: Engine capable of running on a fuel/air mixture significantly lower than stoichiometric.

lean mixture: Inducted air/fuel mixture containing an excess of air.

light-off: Coming into operation of a catalytic converter when operating temperature is reached.

light-off time: Time from engine start to operating temperature of an exhaust catalytic converter.

liquefied natural gas: Natural gas as an automotive fuel, principally consisting of methane.

liquefied petroleum gas: An automotive alternative fuel consisting mainly of propane, sometimes with the addition of butane, which can be stored at relatively low pressure. Also know as LPG.

liquid-cooled: Cooled by conduction and convection to the passage of a liquid, such as water.

lobe: Part of a cam profile that extends beyond the base circle and brings about the lift of a follower or tappet.

loop-scavenging: Scavenging of a two-stroke engine by vertical circulation within the cylinder and to the exhaust port of unburned gases ahead of the incoming charge.

M

main jet: Principal jet in a carburetor through which the greater proportion of fuel flows in normal steady-state operation.

main metering system: Pertaining to a carburetor float chamber, main discharge tube, jet, or venturi as a complete functional unit.

manifold: System of ducts or pipes that divides flow and conducts it to more than one point of delivery, or that unites a flow from a number of sources for delivery at one point.

manifold absolute pressure: The mean gas absolute static pressure in an engine-induction manifold, usually measured by a mercury scale.

manifold depression: Difference between the mean static pressure within the engine induction manifold and the ambient pressure. Although the difference is usually negative in a naturally aspirated engine, it will be positive in a supercharged engine.

map: Electronic data bank from which signals controlling one or more parameters can be outputted in consequence of one or more signals.

maximum brake power: Maximum measured power of an IC engine measured on a dynamometer.

mean effective pressure: The mean pressure in an engine cylinder during the working or power stroke.

mean piston speed: Effective distance traveled by a piston per unit time as a function of crankshaft rpm.

metering rod: Valve consisting of variable-section rod and orifice by which flow of a fluid can be metered.

methane: Hydrocarbon constituent of natural gas, also produced by sludge digestion.

methanol-methyl: Alcohol as an alternative fuel source.

methyl tertiary butyl ether: Oxygen-bearing antiknock fuel blend component known as MTBE.

misfire: Failure of a fuel charge to fire or ignite in the proper way.

mixing chamber: The barrel of a carburetor.

MON: Motor octane number.

monolith: Originally a large block of stone, but used to describe a catalytic converter that is made of a ceramic honeycomb-shaped element.

Morse test: Engine test for multi-cylinder engines in which each cylinder of the running engine is stopped in turn so that its contribution to the total output can be measured.

motor method: A standard test for fuel octane.

Motor octane number: A guide to the antiknock performance of a fuel under relatively severe driving conditions, such as full throttle with elevated engine coolant temperatures.

motoring test: Method of testing an engine without combustion present while being turned over by an electric motor to determine frictional and pumping losses.

multi-point fuel injection: A fuel-injection system in which fuel is administered to each cylinder by a separate fuel injector.

multi-spark ignition: A primary ignition event initiated by a rapid sequence of sparks rather than a single spark.

multi-throw crankshaft: Shaft with cranks set at equal angles of less than 180 degrees.

mushroom valve: A valve with a narrow stem surrounded by a disk, like a head with a concave face.

N

NACA duct: Aerodynamically profiled shallow duct, developed by the United States National Advisory Committee for Aeronautics.

narrow V-engine: Engine of V-configuration with cylinder axis less than 60 degrees to each other, and in some cases at an angle so narrow that the cylinders are staggered to avoid interference.

needle bearing: Rolling-element bearing in which the rollers are small-diameter needle-like elements.

needle valve: Valve by which fluid flow rate is controlled by the degree of insertion of a tapered needle into a fixed-diameter orifice.

net power: Brake power of a fully equipped engine.

nitrile rubber: Synthetic rubber polymer of butane and acrylonitrile, noted for its oil, fuel, and temperature resistance, and therefore often used in shaft seals.

nitromethane: Organic liquid, sometimes used as a fuel, with the addition of methyl alcohol.

nitrobenzene: An organic fuel additive that is highly poisonous.

nitrogen oxides: Compounds of nitrogen and oxygen produced during combustion and conforming to the general formula NOx.

normally aspirated: An engine that breathes air at atmospheric pressure.

O

octane number: Measure of the antiknock properties of a fuel, particularly gasoline, derived from comparative testing using a standard variable-compression engine. The antiknock properties of the fuel under test are compared with those of a fuel containing iso-octane (given an arbitrary rating of 100) and heptane (rated at zero). The volumetric percentage of iso-octane is taken as the octane rating of the fuel under test. Three values exist: motor, research, and front-end octane.

Open-cycle turbine: A gas turbine operating cycle in which air enters from atmosphere and is discharged to the atmosphere along with the byproducts of combustion.

open-loop engine control: Control of engine parameters such as ignition timing and fuel mixture by preset, often mechanical controls.

opposed cylinder engine: Reciprocating engine in which pairs of pistons operate in each cylinder, the combustion chambers being formed between the piston crowns. Also known as a deltic engine.

organic compound: Substance containing carbon and hydrogen, possibly with other elements. Not all organic compounds are necessarily of natural derivation, like coal and natural gas. Some are synthetically produced.

O-ring: A toroidal elastometric static seal.

oxygenates: Alcohol-based fuels or fuel additives.

P

peak brake power: Highest power developed within the speed range of the engine.

peak torque speed: Engine speed at which peak torque occurs.

pentane: Light and volatile paraffinic constituent of gasoline that aids in starting.

pilot jet: Carburetor jet that admits

fuel downstream of the throttle for starting.

pitting and piling: Erosion of one ignition contact breaker contact and deposition of metal on the other resulting from high-temperature vaporization and incorrect condenser capacity.

plenum chamber: Any chamber at which air or gas is held at higher than ambient pressure. This term is modified when dealing with an intake manifold, described as a chamber that connects the runners of an intake manifold and is under a vacuum on a normally aspirated engine.

pollutant: A substance released into the environment that represents a hazard to man or nature.

poppet valve: Disk-shaped valve, with the disk attached to a stem, the reciprocating movement of which causes the valve to open and close.

popping back: Premature ignition of the air/fuel mixture in the intake manifold.

positive crankcase ventilation: Emissions-control device that prevents crankcase gases from entering the atmosphere, usually by drawing the gases from the crankcase and feeding them to the engine to be burned.

positive-displacement pump: A pump that delivers a fixed or metered amount of liquid per stroke or cycle.

post-ignition: Ignition emanating from hot spots within the engine cylinder and occurring after the initiated ignition.

power stroke: Stroke of a reciprocating engine during which the piston moves under the effect of combustion pressure.

preheated catalyst: Exhaust catalytic converter that heats up on or before starting of the engine, and is not therefore delayed in operation by the need to absorb heat from the engine exhaust. Usually electrically preheated.

pre-ignition: Ignition of the air/fuel mixture in a spark ignition engine before the timed spark, caused by a hot glowing surface.

pressure-wave supercharging: Increasing engine intake pressure by tuning the intake duct to resonate at a particular operating speed.

proportional control: Closed-loop engine-control system in which the feedback signal is proportional to the output of the measured parameter.

pulse air injection: System that uses exhaust-pressure pulsations to force air into an exhaust system to oxidize exhaust pollutants.

pumping losses: Part of the total or indicated power that is expended on the induction of the fuel and air charge into an engine, and expelling the gases on the completion of combustion.

Q

quarter wavelength resonator: Acoustic device of cylindrical form with a preset or adjustable resonant frequency determined by length.

quench: The cooling of a portion of the cylinder head gases during combustion in an IC engine, usually by minimizing the clearance between the piston crown and cylinder head. The high surface-to-volume ratio effects rapid heat loss and suppresses detonation, though at the expense of higher hydrocarbon emissions.

quiescent chamber: Type of combustion chamber of a direct-injection diesel engine in which the mixing of the charge is predominantly non-turbulent or quiescent.

quiescent combustion: Non-turbulent combustion, particularly in a lower-speed diesel engine and normally employing a multi-nozzle injector.

R

radial engine: Multi-cylinder engine in which the cylinders radiate from one single-throw crankshaft.

ram air: Air in orderly rapid motion relative to the vehicle and therefore possessing kinetic energy, particularly in context of air required to charge an air-breathing device such as a cooling or engine-induction system.

Ram-air induction: Pressure charging using the momentum of the induction air in an intake pipe of tuned length.

ram duct: In an engine with two inlet valves of specific function, the part of the inlet tract that provides ram air, usually without intentional swirl.

reformulated gasoline: Gasoline formulated to minimize exhaust emissions, having, for example, low vapor pressure and low aromatic and sulfur content, with the addition of oxygenates to reduce emissions of carbon monoxide and hydrocarbons.

Research octane number: A guide to antiknock performance of a fuel under mild driving conditions. It is derived from one of the standard comparative tests using the Cooperative Fuel Research engine.

retarded ignition: Timing of the ignition spark to occur later than the optimum timing for fuel of a specified octane rating.

rich mixture: Fuel/air mixture in which the proportion of fuel exceeds that necessary for the theoretically complete or stoichiometric combustion.

road octane number: Usually the octane number of primary reference fuel that just gives trace knock in a vehicle on the road or on a chassis dynamometer under specified conditions.

RON: Research octane number.

Roots blower: Mechanical-induction pressure charger in which air pressure is provided by contiguous rotation of two or three lobed rotors or pistons.

runaway knock: Engine knock that becomes progressively worse under steady speed and load conditions.

runaway surface ignition: Improper ignition within engine cylinders emanating from hot engine components (not from carbon deposits), becoming more pronounced as engine temperatures increase.

S

SAE rating: The standard for engine horsepower measurement for an engine with no parasitic components attached, such as an alternator, power steering pump, or any other ancillary.

SAE viscosity: The viscosity of motor oil per the standard of the Society of Automotive Engineers.

scavenger: Halogens present in a lead antiknock compound that prevent the buildup of lead oxides and sulfates in an engine's combustion chamber by rendering lead compounds formed during combustion volatile so that they may pass through the exhaust.

scavenging: The removal of exhaust gases from an engine cylinder, particularly by an induced flow of gas.

scavenging efficiency: The ratio of the mass of working gas retained in an engine cylinder at completion of the exhaust cycle to the total charge mass supplied.

Schrader valve: A type of pneumatic non-return valve used especially for tire inflation. Depression of a central spigot opens the valve to facilitate inflation and the measurement of tire pressure. This style valve is also used in other applications.

secondary: When pertaining to engine balance, an event that occurs at twice engine speed.

secondary couple: The out-of-balance disturbance in an engine that causes rocking about a horizontal or vertical axis at right angles to the crankshaft and at twice the engine speed.

secondary venturi: A small venturi mounted coaxially within the main venturi of a carburetor. More commonly known as a booster venturi.

sensitivity: When pertaining to gasoline, the difference between the research and motor octane ratings.

sensor: Device in a control system

that provides information from an input. An example would be an oxygen sensors reading of oxygen content in an engine's exhaust.

side-draft carburetor: Carburetor with a horizontal barrel.

Slide-operated carburetor: Carburetor in which the volume flow of air is regulated by a simple slide valve in the inlet track. Commonly found on motorcycles and light industrial engines.

Slow-running jet: A carburetor jet that compensates for the natural reduction in mixture strength at low engine speed by supplying excess fuel.

sonic throttling: Increasing the flow velocity of a gas or vapor in a pipe, tube, or orifice to sonic speeds by throttling, as applied to carburetion or engine-induction airflow.

spark knock: Detonation within an engine cylinder caused by excessively advanced ignition timing rather than surface ignition.

specific fuel consumption: Fuel consumed per unit of power output, usually expressed as mass or volume per unit power, per unit time, for example, pounds per brake horsepower hour.

square engine: Engine in which the cylinder bore diameter is equal to the piston stroke.

squish: The squeezing of the air/fuel mixture in an engine's combustion chamber away from the end-gas region at the top of the compression stroke. Also the area of the engine combustion chamber with minimal clearance between the piston crown and cylinder head at top dead center.

squish lip: Bowl-in-piston combustion chamber featuring squish with reentrant bowl and hard-edged entrance from the bowl to the crown.

stoichiometric ratio: Ratio, usually of mass, between air and a flammable gas or vapor, at which complete combustion or chemical combination takes place. Often referred to as the chemically correct mixture.

stratified charge: Reciprocating engine-combustion system in which combustion is initiated in a layer of relatively fuel-rich mixture, and then spreads to a much weaker mixture elsewhere in the combustion chamber to increase fuel economy.

supercharge: Artificial increase in pressure of induction air or gases achieved by pressure charging from a mechanical pump.

surface carburetor: Early form of carburetor in which the induction air is passed over an exposed surface of fuel.

surface ignition: Ignition emanating from hot spots within an engine cylinder rather than from a timed spark.

surface/volume ratio: Numerical ratio of surface area to volume, specifically in the combustion chamber of an engine.

surfactant additive: Organic fuel additive that forms a coating on metals or other surfaces, which it protects, acting also as a detergent or dispersant by making deposits partially soluble.

swept volume: In a reciprocating engine, the volume of the cylinder formed by the bore diameter and the stroke of the piston.

swirl: Orderly rotation of combustion gases in an engine cylinder to improve mixing and heat transfer.

swirl chamber: Small chamber or cavity formed in a cylinder head of an engine to promote swirl.

T

tailpipe emissions: Emissions from an engine's tailpipe, as opposed to evaporative and other emissions.

TAME: Tertiary amyl methyl ether. Oxygenate used as a gasoline blend component.

thermal efficiency: Ratio of useful work performed by an engine to the total energy content of the fuel consumed. A measure of efficiency of the combustion process.

thermal loading: Total effect of heat and temperature on mechanical and structural components, particularly of an engine.

thermodynamic cycle: Idealized sequence of operating stages of an engine or other heat machine, often illustrated by a diagram showing pressure plotted against displacement volume. Actual engine cycles of operation differ considerably from their ideals.

throat: Narrow end of a tapered aperture or venturi of a carburetor.

throttle: A butterfly valve for controlling the flow of fuel and air through a carburetor.

tuned intake pressure charging: Increasing the mass of the inducted fuel mixture by matching the acoustic resonance of the induction system to engine speed.

U

unburned hydrocarbons: Unburned or incompletely burned products of engine combustion released into the atmosphere as an exhaust emission.

unleaded gasoline: Gasoline that has no added lead alkyls as octane improvers, but may nevertheless contain a specified very low maximum amount of lead.

updraft carburetor: Carburetor in which the inducted airflows upward past a jet, usually located in a central venturi.

V

vacuum advance: Mechanical-pneumatic system for advancing ignition timing, using carburetor throat depression as a source of vacuum.

vacuum carburetor: Carburetor in which the air slide is controlled by suction or depression from the inlet tract.

valve lifter: Valvetrain component that bears directly on the cam. Also known as a tappet or follower.

valve timing: Geometric positions in relation to the crankshaft rotation or other reference at which valves open and close.

vapor-lock: Interruption of flow in a piped fluid system resulting from vaporization of the fluid or entrapment of air.

vapor-lock index: Index that combines the reed vapor pressure and the percentage evaporated at a specified temperature of a gasoline. It measures the fuel's propensity for vapor-locking.

variable-compression engine: Usually a standard engine for laboratory or research use in which the compression ratio can be varied, particularly while the engine is operating.

venturi: Convergent-divergent nozzle that accelerates and lowers static pressure in gases or vapors flowing through it. In a carburetor, the venturi provides the depression in the airflow pressure that causes fuel to be drawn from its bowl or chamber into the air stream.

volumetric efficiency: The amount that the cylinder of an engine is filled by the incoming charge following an exhaust stroke. A measure of the ability of an engine to breathe freely. Volumetric efficiency is a ratio of masses, not volumes. The original definition was the ratio of the volume of induced charge at the inlet valve temperature and pressure to the engine's swept volume.

W

warmup: The period of engine operation between starting and reaching a stable operating temperature.

wastegate: Valve that diverts exhaust gases away from a turbocharger turbine when charge pressure has reached the requisite figure.

water injection: Injection of water mist directly or indirectly into a cylinder of an IC engine to cool the incoming charge and limit detonation.

wick carburetor: Early form of carburetor in which induction air gathered fuel vapors from an exposed wick.

OTHER BOOKS BY HPBOOKS

HANDBOOKS

Auto Electrical Handbook: 0-89586-238-7/HP1238
Auto Upholstery & Interiors: 1-55788-265-7/HP1265
Car Builder's Handbook: 1-55788-278-9/HP1278
The Lowrider's Handbook: 1-55788-383-1/HP1383
Powerglide Transmission Handbook:1-55788-355-6/HP1355
Street Rodder's Handbook, Rev.: 1-55788-409-9/HP1409
Torqueflite A-727 Transmission Handbook: 1-55788-399-8/HP1399
Turbo Hydramatic 350 Handbook: 0-89586-051-1/HP1051
Welder's Handbook: 1-55788-264-9/HP1264

BODYWORK & PAINTING

Automotive Detailing: 1-55788-288-6/HP1288
Automotive Paint Handbook: 1-55788-291-6/HP1291
Fiberglass & Composite Materials: 1-55788-239-8/HP1239
Metal Fabricator's Handbook: 0-89586-870-9/HP1870
Paint & Body Handbook: 1-55788-082-4/HP1082
Pro Paint & Body: 1-55788-394-7/HP1394
Sheet Metal Handbook: 0-89586-757-5/HP1757

INDUCTION

Bosch Fuel Injection Systems: 1-55788-365-3/HP1365
Holley 4150: 0-89586-047-3/HP1047
Holley Carbs, Manifolds & F.I.: 1-55788-052-2/HP1052
Rochester Carburetors: 0-89586-301-4/HP1301
Turbochargers: 0-89586-135-6/HP1135
Weber Carburetors: 0-89586-377-4/HP1377

PERFORMANCE

Baja Bugs & Buggies: 0-89586-186-0/HP1186
Big-Block Chevy Performance: 1-55788-216-9/HP1216
Big-Block Mopar Performance: 1-55788-302-5/HP1302
Bracket Racing: 1-55788-266-5/HP1266
Brake Systems: 1-55788-281-9/HP1281
Camaro Performance: 1-55788-057-3/HP1057
Chassis Engineering: 1-55788-055-7/HP1055
Chevy Trucks: 1-55788-340-8/HP1340
Ford Windsor Small-Block Performance: 1-55788-323-8/HP1323
4Wheel&Off-Road's Chassis & Suspension: 1-55788-406-4/HP1406
Honda/Acura Engine Performance: 1-55788-384-X/HP1384
High Performance Hardware: 1-55788-304-1/HP1304
How to Hot Rod Big-Block Chevys: 0-912656-04-2/HP104
How to Hot Rod Small-Block Chevys: 0-912656-06-9/HP106
How to Hot Rod Small-Block Mopar Engine Revised: 1-55788-405-6
How to Hot Rod VW Engines: 0-912656-03-4/HP103
How to Make Your Car Handle: 0-912656-46-8/HP146
John Lingenfelter: Modify Small-Block Chevy: 1-55788-238-X/HP1238
LS1/LS6 Small-Block Chevy Performance: 1-55788-407-2/HP1407
Mustang 5.0 Projects: 1-55788-275-4/HP1275
Mustang Performance (Engines): 1-55788-193-6/HP1193
Mustang Performance 2 (Chassis): 1-55788-202-9/HP1202
Mustang Perf. Chassis, Suspension, Driveline Tuning: 1-55788-387-4
Mustang Performance Engine Tuning: 1-55788-387-4/HP1387
1001 High Performance Tech Tips: 1-55788-199-5/HP1199
Performance Ignition Systems: 1-55788-306-8/HP1306
Small-Block Chevy Performance: 1-55788-253-3/HP1253
Small-Block Chevy Engine Buildups: 1-55788-400-5/HP1400
Stock Car Setup Secrets: 1-55788-401-3/HP1401
Tuning Accel/DFI 6.0 Programmable F.I: 1-55788-413-7/HP1413

ENGINE REBUILDING

Engine Builder's Handbook: 1-55788-245-2/HP1245
How to Rebuild Small-Block Chevy LT-1/LT-4: 1-55788-393-9/HP1393
Rebuild Air-Cooled VW Engines: 0-89586-225-5/HP1225
Rebuild Big-Block Chevy Engines: 0-89586-175-5/HP1175
Rebuild Big-Block Ford Engines: 0-89586-070-8/HP1070
Rebuild Big-Block Mopar Engines: 1-55788-190-1/HP1190
Rebuild Ford V-8 Engines: 0-89586-036-8/HP1036
Rebuild GenV/Gen VI Big-Block Chevy: 1-55788-357-2/HP1357
Rebuild Small-Block Chevy Engines: 1-55788-029-8/HP1029
Rebuild Small-Block Ford Engines: 0-912656-89-1/HP189
Rebuild Small-Block Mopar Engines: 0-89586-128-3/HP1128

RESTORATION, MAINTENANCE, REPAIR

Camaro Owner's Handbook ('67–'81): 1-55788-301-7/HP1301
Camaro Restoration Handbook ('67–'81): 0-89586-375-8/HP1375
Classic Car Restorer's Handbook: 1-55788-194-4/HP1194
How to Maintain & Repair Your Jeep: 1-55788-371-8/HP1371
Mustang Restoration Handbook ('64 1/2–'70): 0-89586-402-9/HP1402
Tri-Five Chevy Owner's Handbook ('55–'57): 1-55788-285-1/HP1285

GENERAL REFERENCE

Auto Math Handbook: 1-55788-020-4 /HP1020
Corvette Tech Q&A: 1-55788-376-9/HP1376
Ford Total Performance, 1962–1970: 1-55788-327-0/HP1327
Guide to GM Muscle Cars: 1-55788-003-4/HP1003
The VW Beetle: 1-55788-421-8/HP1421

MARINE

Big-Block Chevy Marine Performance: 1-55788-297-5/HP1297
Small-Block Chevy Marine Performance: 1-55788-317-3/HP1317